Physics in Fours:

Mindful Adventures

George C. Whittemore

TABLE of CONTENTS

Foreword and Dedication..Pages 3-13

Introduction...Pages 14-28

Four Fundamental Concepts:

       Space............................Pages 29-57

       Time..............................Pages 58-89

       Mass..............................Pages 90-104

       Charge...........................Pages 105-149

Four Fundamental Forces:

       Force..............................Pages 150-173

       Gravity...........................Pages 174-195

       Strong Nuclear................Pages 196-212

       Weak Nuclear.................Pages 213-224

       Electromagnetism...........Pages 225-241

Four Conservation Laws:

       Conservation of Linear Momentum..............Pages 242-252

       Conservation of Angular Momentum............Pages 253-261

       Conservation of Charge................................Pages 262-264

       Conservation of Total Energy........................Pages 265-288

Acknowledgements..Pages 289-290

FOREWORD

I began writing this book in 1984, six years into my teaching career. Thirty-six years have passed since I typed those first words, and with that passage of time much has changed. I have taught all levels of physics, starting from my brief stint as a teaching and research assistant at WPI in 1978. In the intervening years, I have worked as a physics consultant for the layout designers at Digital Equipment Corporation, as well as a public-school teacher, independent school teacher, coach, Science Department Chair, Dean of Students, and Headmaster at an urban public high school. At the moment, as a semi-retiree, I have spent the last several years teaching physics to mainly biology and chemistry majors at a small liberal arts college.

During that time, Digital went under, teaching remained teaching no matter the venue, being a Dean of Students was a valuable but often painful experience, running a department had its share of headaches and rewards, and helping to run a large, diverse school was a wild ride indeed. Through it all, my love for teaching and the teaching profession has only grown, despite the best efforts of external "experts" to sap the joy out of the classroom by requiring an endless array of "accountability" measures that serve mainly to *detract* from excellence in the classroom.

Master teachers are virtuoso jazz musicians: constantly improvising within a structure, feeding off the "music" of questions and dialogue to arrive at a deeper understanding of the content, accompanied by a profound sense of wonder at the universe we live in for an entirely too brief a moment in space and time. College teaching has been an absolute blast, and teaching in general is the best career choice I could have made despite the current trend of tasking teachers with mountains of documentation, thereby diverting time and energy away from actually perfecting the science and art of teaching and learning. So, I arrive back at the keyboard with a far deeper understanding of teaching, textbooks, students, learning and life.

Emerging from graduate school, and many times since then, I have been constantly reminded about the anxiety most people feel regarding the study of physics. I am still hoping for the day when a conversation concerning what I do for a living goes something like this:

"What do you teach?"

"I teach physics."

"Really? Wow, that's great – it's such a creative, vibrant, imaginative field of study!"

These days what I usually get is more like this:

"Oh, *physics*. I flunked that in high school. That stuff is way over my head!"

Why is the latter reaction far more prevalent than the former? That certainly has been the case in my experience, and if anecdotal evidence counts for anything, it is a shared experience amongst many of my colleagues. Historians, philosophers, and anthropologists could perhaps enlighten us as to the many reasons why this particular strain of insecurity is embedded into our national psyche and culture. People often act as though it's cool to feign ignorance about science and math, or have stereotypes of physicists, thinking they can't also be artists or athletes or "normal". It's a common fallacy: find the example that fits a stereotype, then extrapolate to an entire population. Of course, all of this leads to a lot of deeper questions, ones I hope to answer in this book, explicitly and otherwise.

I sometimes wander back in time to when we were kids. We asked such great questions about the world. Why is the sky blue? How did all of existence come to be? What produces ocean waves? What was there at the start of the universe? How do we know what we think we know? Hundreds of thoughtful, fundamental questions: where does that innate curiosity go? I think some of that creative spark we all have gets snuffed out the first time we pick up the typical physics textbook: huge, occasionally boring, and frequently inaccessible (financially and otherwise) to the average reader. Worse still, all the fun, the new frontier concepts such as the detection of gravity waves, are usually found at the end of a 1500-page tome, if at all! Why? As a consequence, the typical introductory course syllabus never gets anywhere close to those pages, assuming readers are not comatose by the time they hit page 1000, and students miss out on some extraordinary material.

Wolfgang Pauli, a physicist famous for many things, including providing an explanation for why we don't fall through a "solid" floor made up of atoms that are mostly empty space, reportedly once said of a student's work: "that's not even wrong!" To borrow from Shakespeare and throw in a type of eponym, "brevity is the soul of Whit"! *Textbooks are much too long.* This venture will not pass 300 pages, and fun, fascinating, new developments in physics will be interspersed throughout. By keeping our focus on four fundamental concepts in physics, the four fundamental forces that govern all interactions, and the four major conservation laws that *always* hold, we will venture into terrain that offers views across the entire landscape of physics. Now you understand the title of this book!

There are some very well written textbooks and paperbacks about physics out there, but the textbooks of necessity often leave out the rich cultural, historical, and conceptual development so essential for a thorough understanding of physics as a body of knowledge and as a human enterprise. And the paperbacks often lack the rigor required for a serious study of physics. Why can't we have our cake and eat it, too? I started writing this as an ancillary textbook and it ended up like a novel. The thought has occurred to me that this is yet another bi-polar, false dichotomy. Why can't a physics book be both?

Most, if not all, of the material in this book stems from my classroom teaching in various ways. The myriad approaches, analogies, lectures, and demos have worked, if I am to believe the letters from former students and their promise to buy my book, if I ever found the time to actually get it done! A physics book should cultivate that inner desire to learn about things, enable the reader to understand how the universe works the way it does in a sophisticated manner, and all the while tap into the reader's fascination with life and existence itself.

There are six main goals I have in writing this book. The first of these is to provide a comprehensive understanding of the fundamental workings of nature with the same informal yet precise style that I use to teach physics. The intention is to do this in a concise, powerful manner that is accessible to a wide audience and that does not involve a $300, 1500-page tome to carry around in a fortified backpack. There is minimal mathematical emphasis throughout the book, but some ideas are expressed in mathematical form where failure to do so would undermine the reader's understanding of the concept involved. Mathematics is the language of physics in many ways, but formulas and the like are far from the whole enchilada. Also, there are many forms a rigorous approach can take: diagrams, pictures, graphs, allegories, analogies, and so on.

Equations sometimes obscure the content of physics; I will provide some ideas about analyzing formulas in the book with the aim at reducing some of that obscurity.

The second purpose in writing this physics "novel" is to open up a dialogue *about* science. Perhaps this goal is the one closest to my heart. I want to eliminate the misconception of a cut and dry scientific method being applied in endlessly monotonous ways to mundane problems. I want people to see that science is also an art, and a scientist creates ideas regarding nature's rules in much the same way as an artist imagines a painting onto a canvas. It took me a full year to convince an extremely gifted artist in my physics class that we arbitrarily cleave the unity of knowledge into "disciplines" such as art and science, but the reality is that all creations of the human mind are paths that explore the mystery of existence. It is my hope this physics book will have as much success.

In any scientific field, ideas must be tested, repeatedly, in our voyage toward truth. This distinguishes science from many other endeavors, and it makes the adventure all the more compelling. The fact that this process, on average, has resulted in *doubling* my lifespan over the past few centuries (as well as the lifespan of billions of other human beings) is equally as compelling. I read a little while ago that since the earliest fossil records, fully 99% of species that have existed on Earth are now extinct. That should give us all pause. One fact I continually stress to my students is that the Earth does not need our protection – it is not fragile, *we are fragile* as a species. Several Native American tribes know this fact to its very core, and therefore regard Mother Earth and all its creatures with a reverence and a wisdom that seeks to preserve the quality of life for many generations, not just the present moment. The planet is going to be just fine for billions of years until it is swallowed by the Sun as it transitions into a red giant, assuming there is no huge asteroid or black hole that wanders too close to our orbit in the interim.

    I introduce these thoughts because there are many paths through the mountains and up to a summit, yet the methods of science have proven to be extremely powerful, though often painstakingly slow. We live in a most dangerous time, when objective reality, facts, and truth itself are under siege. Our species has traveled this road before, too many times, through the Dark Ages, world wars, famines, disease, and hunger. The power of science comes from the fact that you can't bluff for long – nature will always win in the end and has no regard for opinion or ego. Therefore, while you may have the *opinion* that a match thrown into a gas can is harmless, the facts prove otherwise. **The laws embedded in nature are not subjective beliefs!** History has tried to teach us that every time we reject facts and evidence, people die. And science also teaches us, in the words of Oliver Cromwell and as quoted in *The Ascent of Man* by Jacob Bronowski: "I beseech thee in the bowels of Christ – pray that you may be mistaken". If the evidence proves one's ideas wrong, no matter how beautiful or persuasive, then the ideas are wrong and we search for another vista that will expose patterns in our natural world. This requires immense humility, a quality sorely lacking in far too many "leaders" today.

    I would also like to dispel the stereotypical image of the awkward, introverted physicist who absent-mindedly walks into walls while pondering equations. Yes, that physicist does exist out there and I have personally walked into walls on occasion, but the variety of human beings engaged in the study of physics is vast. We need to talk about the societal ramifications of the work scientists do and the responsibilities a scientist has with regard to ongoing research. I created a course a while back called "Science, Society, and Technology" that investigated many of the controversial, complex, and technical issues facing our citizens today: nuclear weapons and nuclear power, toxic chemicals, euthanasia and advancing medical technology, the use of animals in research, climate change, and so on. Students debated and discussed these issues openly, but only after gaining a firm foothold on the body of knowledge we currently have regarding that issue. We need to know the science involved, the thinking of the courts when these matters go to trial (and they often reach the Supreme Court), and the lessons learned from past practice if such precedent exists. A talented colleague taught other sections of the class and reported the same reactions: students loved this experience because they had a chance to get involved, speak their minds, and get into issues that directly affect our lives. The classes were oversubscribed in a very short time, and the need for this sort of education has never been more critical as we muddle our way through the Coronavirus Pandemic.

For a reasonably advanced technological society, we are too ignorant (in some cases we pass right over ignorance and enter the land of antagonistic) about the workings of science. We need to understand its successes and its failures, and we need to grasp the power of scientific thought and practice while simultaneously cultivating the wisdom to see what science can never achieve for us collectively. Science cannot solve all our problems; such unmitigated faith is unwarranted. Despite all the biological evidence that refutes genetic superiority in any part of our species, racism persists. A chemical in our skin that responds to sunlight leads to…hatred…murder…wars. It makes no sense scientifically; it is also the path of the spiritually blind. Evolution may be the grand plan of a Creator, or it may just be the natural unfolding of carbon based existence. Either could be true, but the colossal amount of evidence in favor of evolutionary science stands by itself, regardless of belief. It could be true that God, Christian or otherwise, made all things, but teaching this as *science* is a lie. It cannot be proven one way or the other; it is a matter of faith, and the most profoundly religious and spiritual people I have known in this world carry that faith with them privately, humanely, and with immense compassion and humility.

We often fall into sophistry; all of us are vulnerable in that regard. Some people claim that prayer is not allowed in public schools. This is false. My students prayed all the time, especially before the physics final exam! What this democracy does not allow is government sanctioned prayer over the PA system that all must participate in, and I am incredibly thankful to our founders that their wisdom, in this realm at least, has prevailed. I ask you to imagine how you would respond if the government forced you or your child to pray in a religion you don't believe in, or forced you or your child to pray *at all*. It is said we live for an extremely short time between two eternities. I have found that religion and science can complement one another in profound ways; both are beautiful expressions of our common humanity. This second goal, exploring the actual nature of science, goes to the very essence of our national psyche, our democracy, and our survival as a species.

The third aim of the novel is to provide readers with an overall view of the present state of knowledge in physics and where the future frontiers lie. In this age of information and unfettered specialization, it is crucial that people have a general picture of where we are now, where we have come from, and where we are likely to go in the near future. Not only does this make good sense in a technology-based society, but it also will provide readers with a healthy dose of how we know what we know, as well as providing direct evidence that we are far from knowing it all! As mentioned previously, nearly all the textbooks I have come across while teaching and learning physics save these fascinating nuggets until page 1000 or so, and coincidentally this is nearly always a page no-one reaches during the course of a semester or even a year of study. Invariably it is precisely these advances in the frontiers of physics that students are absolutely fascinated by and can't get enough of, but sadly are not addressed before we run out of time. The story does not have to unfold in that manner: advances at the leading edge of physics should be interwoven into every page we write! At the outset, the chapter on "Space" will bring this goal to life, and in all the subsequent chapters practical applications of the physics concepts will be discussed along with frontiers in research.

The fourth goal I have in writing this book is to instill in the reader an excitement and an appreciation for the rich history of physics and for the people behind the ideas that changed our way of life. Most textbooks do not even attempt to present the historical backdrop to the progression of ideas and theories in physics. This book is not intended to be a comprehensive guide to the history of physics; those books are out there and I will cite some of the better ones that I have come across throughout this book. Rather, the intention is to reproduce some of the drama inherent in the evolution of ideas about how the universe we live in works, and how the people who came up with these ideas did so, at least as far as we can tell from delving through historical documents. I hope the reader will come to see physics as one of the most creative human endeavors one can undertake, and that the fuel for progress in physics is our collective imagination.

The fifth aim of this book centers on a particularly curious thing about our educational system that I and many others have noted over the years: we seem to value tests so highly but seldom bother to show students the best way to prepare for them and take them. In a broader sense, people need to learn *how* to learn. How does one tackle a certain type of problem in physics? How should one study for a big test or a small quiz? What methods of teaching work well in the classroom? What are the best "habits of mind" that students should cultivate to achieve better success and deeper understanding? How does one read a formal, rigorous scientific textbook? What are the best study methods and habits? I think we are getting better at addressing this idea of teaching students how to learn, but too often I have observed this critical piece missing in our classrooms and in our textbooks.

The sixth and final aim of this novel flows fairly naturally from the last one, and that is to provide teachers, students, and all readers with "best of" examples I have found effective in teaching physics and that involve analogies, stories, and Socratic lectures garnered from many years of teaching in a wide variety of settings and across diverse audiences. Many of these analogies spring from a desire to fully address the insightful and provocative questions students have raised over the years. Any profession that wants to lay claim to being a profession, and I think teaching most definitely is a profession, must share best practices in a similar manner as is done in engineering, law, and medicine. Contrary to popular belief, there is a body of knowledge in the world that has been demonstrated as master teaching. Perhaps even more contrary to common notions about teaching, it takes a great deal of expertise and reflection to teach well, regardless of the subject involved. Too many people think anyone can teach, when they must know from personal experience that such a notion is completely false. Actually, at a superficial level, it is true that anyone can teach. But that is true of any enterprise! Anyone can build a bridge – would you want to cross it if some random bloke designed and built it? Anyone can do surgery – would you want to be the patient if Joe Schmo off the street held the scalpel? Teaching is no different; we have all had master teachers and coaches. Typically, these masters dramatically change lives while constantly transforming their own.

A formal dedication is difficult because there are too many people to include: family, colleagues, Feynman, Einstein, Newton, and every giant known and unknown in the progression of human knowledge. All in some way led me to this writing. Still, I suppose it is natural and inevitable that my students, past, present, and future, figure most prominently in the writing of this physics "novel". Without them, my knowledge and appreciation of physics and of life would be vastly diminished, and this book would never have happened. It has been said or written in many ways, that the goal of life is to find one's gift, the work of life is to perfect that gift, and the meaning of life is to give that gift away, with the goal of improving everyone's life and leaving the world a better place in some fashion. Perhaps this book will, in some fashion, cause a few readers to come into first contact with finding one's gift in life. And so, I raise my glass to all my students and hope they are living this creed, whether or not they actually buy and read this book!

Sláinte!!

Dedication:

In mathematical terms: $4\Sigma S$

{For the sum of all my students}

INTRODUCTION

The following excerpts are from a speech given by Albert Einstein in April, 1918, in honor of Max Planck's 60th birthday. It is one of the best descriptions regarding what draws certain people to science that I have ever read:

"In the temple of science are many mansions, and various indeed are they that dwell therein and the motives that have led them thither. Many take to science out of a joyful sense of superior intellectual power; science is their own special sport to which they look for vivid experience and the satisfaction of ambition; many others are to be found in the temple who have offered the products of their brains on this altar for purely utilitarian purposes. Were an angel of the Lord to come and drive all the people belonging to these two categories out of the temple, the assemblage would be seriously depleted, but there would still be some men, of both present and past times, left inside. ...But of one thing I feel sure: if the types we have just expelled were the only types there were, the temple would never have come to be, any more than a forest can grow which consists of nothing but creepers. .... Now let us have another look at those who have found favor with the angel. Most of them are somewhat odd, uncommunicative, solitary fellows, really less like each other, in spite of these common characteristics, than the hosts of the rejected. What has brought them to the temple? That is a

difficult question and no single answer will cover it. ...A finely tempered nature longs to escape from personal life into the world of objective perception and thought; this desire may be compared with the townsman's irresistible longing to escape from his noisy, cramped surroundings into the silence of high mountains, where the eye ranges freely through the still, pure air and fondly traces out the restful contours apparently built for eternity."

The great physicist, Richard Feynman, compares physics to a colossal game of chess. Nature makes the moves, and through observation, imagination, thinking, and testing, we try to guess the rules of the game that govern those moves. We do this by looking for patterns and symmetries in nature. For example, in a random chess game, if we take the time to gather large amounts of data, we might see a pattern whereby bishops can only move diagonally, and hence discover one of the rules of the game. There is another, much simpler game, called "Petals Around the Rose". It is essentially a puzzle involving six dice. Each roll of the six dice is recorded, and a result is given that is determined from employing a simple rule that governs the game. Here is some sample, random data from a series of dice rolls:

1-3-5-4-6-2 ... yields a result = 6   2-4-6-3-4-1 ... yields a result = 2

3-3-5-6-4-4 ... yields a result = 8   4-2-1-6-6-2 ... yields a result = 0

5-1-5-5-3-4 ... yields a result = 14   6-1-4-3-5-5 ... yields a result = 10

What is the rule for this game that produces these results? Look for patterns in the data. I will tell you that the order of the dice in each roll does not matter; the results would be the same regardless of where each die is placed. Notice that all the results are either zero or an even number. I will tell you that given an unlimited set of data, this pattern or generalization will always hold: all results will be even numbers or zero. One can see that each pattern discerned gives us information, a road to travel down that might yield the solution to the puzzle. I will also tell you that there is more than one way to express the rule that governs the game. This is often the case in physics as well! When one looks for the rule or rules that explain patterns in our natural world, *that* is what doing physics looks like. One final hint regarding this puzzle: the title of the game can be very useful in solving the riddle, but it is not necessary in order to find the solution. Have at it!

All human beings can be scientists, artists, thinkers, and all at once. We are in this grand game together, pondering the great mystery of our own existence as we sail along through space and time, stuck to a small planet in an average galaxy, as part of a vast universe with many billions of galaxies, seen and unseen. We look for patterns and attempt to deduce rules; sometimes we invent rules and then look to see if the data either supports or refutes the theory. Occasionally our imaginations outstrip the ability of technology to test the ideas, and we have to wait for the experimental science to catch up to the theoretical predictions. "Thought experiments", or "gedanken experiments", like those proposed by Einstein which led to huge breakthroughs in our understanding, may appear impractical on the surface, but when looked at deeply, they can revolutionize the way we think and the way we live. At a very young age he asked himself: "What would the world look like if I could travel alongside a beam of light?" That is a most interesting question, and one that leads to some amazing answers which we will soon encounter in subsequent sections of this book.

If we think of the sciences as a giant pyramid, physics would be at its base, supporting all the others above it. As one ascends the pyramid, the degree of complexity of the systems involved increases. Directly above physics is the field of chemistry. Recall that nature is inherently unified and interdependent – it is only humans who divide it into arbitrary disciplines.

Still, there are distinct paths that a chemist takes which diverge (at least momentarily) from those a physicist might follow. Chemistry is a field rife with exceptions to various rules, a fact that can be extremely frustrating for those who major in it! It is more complex than physics in the sense that there are more variables to consider, causing a heightened degree of difficulty in predicting outcomes. Higher up the pyramid is biology, and its vocabulary alone makes it inaccessible to a large portion of the population. The complexity of living systems is remarkably challenging, despite the revolutionary strides that have occurred in the field over the past fifty years or so. Cancer remains a riddle to be solved, as is the case for many other diseases, though millions of lives have been saved nonetheless due to the inspiring progress that has been made through meticulous study and experimentation. Perhaps highest on the pyramid rests geology and environmental science – the study of Earth's systems, such as the weather. The number of variables embedded in these systems is astronomical, which might explain why it is impossible to predict the weather beyond ten days or so with a respectable degree of certainty. Earthquakes, volcanoes, weather systems, and the like tend to be chaotic in nature, and highly dependent on slight variations in initial conditions. Nonetheless, huge progress has been made in the study of climate change (*not* the same as predicting the weather in the short term), and the survival of our species may depend on it. Once again, nature cares not one bit about our beliefs or our politics. When we ignore her patterns, and especially the anomalies in those patterns, we do so foolishly and at the peril of loved ones living now and in the future.

People often cast physicists as the "high priests and priestesses" of science, probably because of the mathematics involved. In my opinion, it is the geologists that should be in that rarified atmosphere! Psychology and sociology are classified as sciences by some, not by others. Perhaps this is a matter of semantics only, since both fields involve experimentation and scientific methodology. Predicting human behavior, either as individuals or as groups in society, is nearly impossible. Yet the study of the human brain, and the attempts to understand societal woes such as racism, may yield solutions that are wedded to the underpinnings of science. When people fully grasp that the color of one's skin is essentially a biochemical process necessary for the survival and adaptation of the human species to particular environments, then we might be in a far better position to grapple with the horrid history of how a false, warped sense of race, with no basis in science and truth, held us in its evil grasp for such a long time and caused so much injustice and death.

One of the greatest mysteries in science is the language it uses, especially in physics, to undergird the entire pyramid: mathematics. Math is not a science, since its validity is generated internally through its own set of logic and rules that do not depend on experimentation. Why does math work so well to explain nature? Nobody knows! The universe we live in is deeply, fundamentally, immersed in the mathematics of geometry, algebra, numbers, calculus, trigonometry, and much more. I have no special wisdom to impart here, other than the fact that math works! Let's take a look at some special numbers as examples of the power of mathematics:

"pi" or π ... many people will know π to be the ratio, *on a flat surface*, of the circumference of a circle divided by its diameter. Any size circle will yield the same number. Fascinating? Perhaps far fewer people know that π can be generated by an infinite, converging sequence of numbers: $\pi = 4(1 - 1/3 + 1/5 - 1/7 + 1/9 - 1/11...)$ ... yields the same result as circumference ÷ diameter! Even more miraculous, π shows up in the formulas describing all types of physical processes, even atomic spectral lines!

The natural log "e" ... probably less familiar to the non-scientist or engineer, yet it also appears throughout nature, in probability, and in many exponential processes. It also can be generated by a sequence:

e = 1/1 + 1/1 + 1/2 + 1/6 + 1/24 + ...) ... yields 2.718281828... This can be generalized to:

$e^x = \sum (x^n)/n!$ ... where n goes from zero to infinity. This function has the unique property that its derivative gives you back the same function! "e" is fascinating in many ways!

The number zero, or "0", was invented in Mesopotamia and then the Mayans (independently).

The counting numbers 1, 2, 3, and so forth had been around in India a long time before the concept of zero took hold. The idea of "nothing" as a number revolutionized mathematics!

It is common knowledge that the square root of 9 is ±3, the square root of 16 = ±4, and so on. It wasn't long before someone thought about the idea of taking the square root of negative numbers. What number, times itself, yields -1, for example? Ancient Greeks had ideas about this, and of course there is no *real* number that satisfies this equation. **So, imaginary numbers were posited, such that the square root of -1 = "i".** Rules were made for imaginary numbers, just as they are for real numbers. Surprise: imaginary numbers are used throughout physics and engineering to describe actual, physical processes. They work!

Why did I choose these numbers, set in boldface? Because it leads to what many consider as the most beautiful equation in all of mathematics, the Euler identity: $e^{i\pi} + 1 = 0$. This is a truly extraordinary result. It is a Rembrandt, a Picasso, or the music of Mozart. It is the universe beckoning to us in silent eloquence, asking us to unravel mysteries created from the beginning of time itself. How can something that seems so abstract accurately describe things in nature that we sense every day? An experienced and wise teacher knows when to be silent, when to observe, and when to listen. And so I leave the rest of this topic for the reader to ponder and investigate.

How does one study and learn physics? Any concept in physics, to be fully understood, should be taught in three ways: ideas into words, words into equations, equations into graphs or diagrams. If the concept is understood to a considerable depth, a person will be able to move within and about these three realms with relative ease. Too often, we skip right over the ideas and words, and jump to equations and graphs, leaving the learner at a loss to explain the actual concepts being discussed. How else can I explain the fact that while in graduate school, many students could "get the answer" to a problem, but very few *knew what the answer meant*? And since they didn't know what the answer meant, when any change was made in the initial conditions so that the problem-solving process was altered, they were at a loss. I know this because too often I was one of those students.

One of the best introductions to quantum physics, a field even the most dedicated physicists frequently find as too abstract and intractable, is to be found in the third volume of *The Feynman Lectures on Physics*. He takes great pains to develop the concepts, *the ideas*, behind the physics involved before getting into the mathematics. The result is a far deeper understanding of the quantum worlds we live in, even if some of the math introduced later on is puzzling. The first thing that must occur in the study of physics is an understanding of the concepts involved *in words*, before any equation or graph hits the blackboard or crosses the optic nerve. Whether you are studying physics for a class, a major exam like the MCAT, or for sheer pleasure (in which case may the winds of good fortune blow forever in your direction), there are many successful methods and approaches that yield great benefit despite the variations which inevitably arise when humans are involved. The last section of this introduction will outline the commonalities among these approaches – those items which are indispensable and serve as fundamentals of learning and teaching, since the two are in a constant improvisational dance with one another.

The typical science textbook, physics or otherwise, cannot be read like a novel. For any given chapter, if a summary is provided, then read that first. Check out the captions, illustrations, and section headings. Review the questions at the end of the chapter first, before doing any problems. If you can do the questions and problems at the end of the chapter after perusing the material, or perhaps in conjunction with other means of exposure to the physics involved such as class lectures or online sources, then you are done. Go play golf! If you are stuck on answering the questions provided, then you are among 99% of the population and you have accomplished the first giant step in learning: identifying precisely what you do *not* know. Now you can read the textbook with a specific purpose: finding out the answer to the question(s) you have and eliminating that particular confusion.

I have never been a big fan of highlighting textbooks; I find myself absent-mindedly underlining whole sections of text so that I'd be better off highlighting what I don't want to remember! Still, if it works for you then by all means sally forth. Whatever method you choose, textbooks are not to be read page by page in order, and there is no theme or plot, other than to drive the reader bonkers at times. Make sure to write down your questions and ideas while the reading is still fresh in your mind. Be precise and specific. If someone says to me "I'm totally lost and I don't understand anything", it's very difficult to help this person because there's no place to begin. If another person says to me "I don't understand why there is no work done on an object in uniform circular motion", I am all in! I know this person has done some work of their own, and I know exactly how to help them learn the physics.

Classroom learning, taking notes, and things of that ilk are often more challenging than reading a textbook. Some of that depends on the expertise of the delivery, yet being thorough and organized goes a long way toward helping to review the material at a later date. With today's technology, it is possible for students to take snapshots of board work, and I am fine with that practice, with the caveat that those photos should be copied into a notebook soon thereafter. I have found that copying notes over, if done in an atmosphere of no distractions, reaps the dual benefits of assimilating much of the material again while also yielding organized, neat notes for future reference. On the flip side, the teaching process should be well structured and involve a fair amount of story-telling with a purpose, practical applications, topics that relate to the interests of the students, humor, and a type of Socratic dialogue that cultivates an atmosphere that welcomes relevant questions. All of these qualities require thoughtful preparation within a framework that allows for improvisation. Master teaching is similar to what an accomplished jazz musician experiences: an ongoing, complex, and creative process that constantly innovates within a structure that feeds off the inputs of others. This concept extends nicely into nearly any endeavor. Imagine seeing a beautifully orchestrated fast break in basketball, complete with intricate and precise passing that involves every player on the floor, capped off by a stunning slam dunk that gets the crowd on its feet.

One of the best books I have read regarding the art and science of teaching is titled: *A Pedagogy for Liberation,* by Ira Shor in collaboration with Paulo Freire. There is a passage in the book that posits teaching as an exercise in perpetual discretion. The complexity arises from the fact that thousands of decisions are being made, on the fly, and often within very short time frames. That is the reason every outstanding teacher, when they hear things like "the five methods of excellent teaching", may listen to the offering but with a fair degree of skepticism born from the intuitive experience of knowing *there are thousands of effective teaching methods available for any given moment in the learning process.* The genius of the master teacher is finding the right method, or perhaps inventing a new one, that suits the moment and the individual learning style while maintaining rhythm and pace. And when this happens, people feel the magic, the art, the music, and the experience is transformative.

The only way to effectively study physics, whether for an exam or for some other purpose, is to *do physics*! By that I mean: do problems, do labs, try to answer questions; in other words, jump in the pool if you're going to learn to swim. Excellent coaches know this to be true and employ it in every practice session. I can diagram and describe the perfect foul shooting technique until the cows come home, but unless players actually take the foul shot and practice proper methods of shooting, it is all for naught. I often ask students how they have prepared for an exam, and too often the response is something like "I read over my notes". Reviewing notes is necessary, but it is not sufficient. Try the problems! Identify the things that are still confusing or that cause a snag in working toward solutions, then take the steps necessary to resolve those issues. And that is a perfect segue for an initial foray into the fine art of problem solving.

In letters from many former students, the feedback I have received over and over is the fact that learning to problem solve was one of the most valuable skills gained from taking physics. Formulas and technical jargon may be forgotten over time unless there is repeated usage; methods of problem solving can last a lifetime. Regardless of the field of endeavor one chooses, problem solving will be an indispensable component of any career and in all of life. I think we can do better in terms of teaching the skills and methods of problem solving. There is so much emphasis on exams, test scores, and letter grades. Why is there not a commensurate emphasis on effective problem solving so that people can have a better chance at excelling?

The following is a suggested sequence for problem solving in physics, though much of it can be extrapolated to other fields and different venues. The intention here is not, however, to short-circuit creative approaches that either alter the sequence or leap outside it wholesale.

1. Read the problem carefully. Understand the goal of the problem. Look for hidden, tacit, and/or irrelevant information. Note the assumptions implicit in the problem.

2. Draw a clearly labeled and legible diagram or illustration depicting the salient aspects of the problem. Keep it simple; this is not a Picasso, but don't let it resemble a Pollock!

3. Write down what is known or given in the problem. Retain proper units for anything given, and read again, looking for words which translate into known variables. In this step, proper nomenclature should be used so that when equations are employed, the information will be substituted into the correct places!

4. Identify precisely which unknowns or what the desired result is to be for the problem. Pay attention to the proper units for the solutions you are seeking, as sometimes dimensional analysis can bring you to the right answer even if you are missing some concepts.

5. Identify the branches of physics being addressed: is this a combination of kinematics and dynamics? What areas of study are relevant to the problem? From there, identify the key equations and concepts that pertain to those areas. At least one of the equations or concepts should include the unknowns you are looking for!

6. Solve for the unknowns in terms of the algebraic symbols first – no numbers yet!

7. Check your solution(s) for proper units. If the units don't check out, mistakes have been made and need to be addressed.

8. Check your solutions for extreme cases – things should reduce to expected results in those instances! If not, regroup and check your math.

9. Now put in known values and variables, if the problem asks for explicit numerical answers, and solve the equations for the solutions.

10. Check your answers for "reasonableness". Do the answers "make sense"? If you have the car going faster than the speed of light, then you need to either check your math and your work or figure out how to manufacture such a machine! Indicate your answers clearly.

There are many pitfalls to avoid in problem solving, and I will offer several of the most common ones, as well as one that is often overlooked. First, too many distractions can cause a lack of focus. Any problem worth solving requires your undivided attention. Second, lack of proper preparation in the form of organized notes or effective study habits will either delay or completely block your paths to possible solutions. Third, understand the physiology of the human brain, at least as much as we know now. Though our knowledge regarding how the mind works has grown enormously in the past few decades, it still remains a mystery in profound ways. For most people, studying, reading, or problem solving in short half-hour spurts is more effective than slogging through hours of anguish, much of which might involve staring at the page. It often happens that while trying to solve a challenging crossword puzzle, the act of simply stepping away for a time, or even taking a nap, sets up the "miracle of return", when the answer pops out at you immediately once you come back to the material with a fresh perspective and renewed vigor. Fourth, beware of anxiety and lack of confidence. They often work in tandem, and if allowed to fester, will prove to be tenacious roadblocks toward finding solutions.

The last pitfall I encountered a very long time ago while reading one of the best books on philosophy and life I have ever come across: *Zen and the Art of Motorcycle Maintenance* by Robert Pirsig. In this passage, he describes how people often get stopped by a problem because of "value rigidity", and he writes of a coconut trap to illustrate the meaning of this potential pitfall. This coconut trap is used to capture monkeys, and before going further I want to be sure that if such a trap is used, then the only moral or ethical reason for it would be to treat an injured animal or to somehow help preserve the species by relocating them to a more hospitable habitat in the wild. Let us hope this is how such traps will be employed. This is how the trap works: a small hole is made in the side of the coconut, but it is a hole large enough for the monkey to reach in and grab the rice that is placed inside the trap. Now the monkey has the rice in his grasp, but the hole is too small for his clenched hand holding the rice to be rid of the trap. He can't escape unless he lets go of the rice! His freedom, momentarily at least, rests solely on the ability to let go! And this is exactly what happens to us with surprising frequency when solving problems. We grab hold of an attractive (we think) solution, we refuse to let go, and we remain trapped. The caution Pirsig suggests in order to avoid this trap, or at the very least to escape from it, is a very wise one: *slow down*! Reevaluate what you deemed as important. Allow your creative sensibilities to flourish; if you are not making progress, seek alternative paths. Let go!

Every time a question is asked, imagine a dot being placed on a blank sheet of paper. As we proceed toward the answer, a circle starts to form with the dot as its starting point. When the answer is found, the circle is complete and we are back where we started at the original dot, but now imagine the area enclosed by that circle is the knowledge gained in the process. The deeper the question, the bigger the circle becomes, along with the knowledge gained. To increase our understanding of nature, we often ask more questions and complete other circles which may or may not intersect or include the first. Where all these questions come from – experience, dialogue, imagination – is a fascinating question by itself. It is critically important to emphasize again that for any law, theory, or hypothesis in physics to form a true circle, it must pass the ultimate test: *repeated results from experimentation.*

Readers should be aware that theories and models are approximations of reality, and therefore are transient in the sense that they are *not* the final word, regardless of the name or the great minds that may have produced the theory. History has repeatedly shown that every once in a while, somebody – usually a person good at asking questions and stubborn enough to find complete answers – comes along and gives a better approximation of reality. This new circle of knowledge will encompass the older, mainly successful theory, while also explaining and predicting new phenomena. Maybe someday *you* will show us how to unify gravity with quantum theory, or discover the true nature and origin of dark matter. Then, when you're rich and famous, you may remember your physics teachers and smile down upon them!

It is my hope that this book will raise more questions than it answers, and that the reader will take the time to write down those questions, thoughts, and ideas that arise throughout. In our culture, we seem to want to rush through everything as if what we are doing at the present moment is a waste of time – the real action, what we really want to do is...somewhere else. I recall several occasions, while dining out in beautiful restaurants in the city, seeing friends congregating around a table, all on their phones! They are right there with each other, poised to share stories and laughter, but the action apparently is "somewhere else". Social media interactions are a poor substitute for actual living, sharing, and being in the present moment. At least that is how I see it.

Many of us have been subliminally coerced into believing our thoughts, our ideas, our conversations within the mind are somehow insignificant compared to someone or something online or on television. People can shrink into a shell, not wanting to appear "stupid" by asking a question that is probably on everyone's mind and is actually quite smart. As a teacher, it is my job to be acutely aware of that fact if it is occurring, and to explicitly convey a culture of welcoming questions and momentary confusion. Yet far too often we plow forward, burying all those questions which can serve to illuminate the understanding of *everyone* in the room or hall, not just the person asking the question. In my experience, the questions people ask are almost always significant and are, at times, brilliant. We must listen to our own ideas and each other. I'm asking the inventor and creator trapped in each one of us to rise from doubt and complacency and do justice to the collective genius which is ours. Listen to *your* ideas instead of the "experts" – let them come in last for a while!

Four Concepts: Space, Time, Mass, and Charge

The four concepts of space, time, mass, and charge cover a vast array of physical phenomena. In fact, nearly any idea in physics can be traced back to one or more of these fundamental concepts. This chapter delves deeply into the idea of "space", and in the process, will challenge common notions of what it is (and is not), as well as how it is measured and operationalized in physics.

SPACE

What is meant by "space" and how is it defined in physics? Some may think of vast regions of black emptiness reaching out beyond the Earth's life-sustaining atmosphere. But there is no border between space and "non-space". The blueness of our world melts gradually into black as we travel outward away from the surface. Perhaps you have heard the conundrum that goes like this: "the universe is not in space, it is space that exists within the universe". You might think that this is a type of koan, like imagining the sound of one hand clapping. It could be construed as such, because there is a mental picture in many minds of a static universe occupying a pre-determined space. We have found such a construction to be false.

The concept of space has several definitions, either to be found in dictionaries or in the everyday parlance of "I need some space". We take the vastness of the universe in all directions and give it the label "space". Labels and definitions can be useful as a means of communication, provided all parties agree on the definition. We often tacitly assume that by defining something, or by giving it a label, we now understand it better and the story is finished. Typically, that is not the case. For example, think about the question: why do objects fall to the Earth when released? Once properly schooled, everyone yells "GRAVITY" and moves on to supposedly deeper issues. But the word "gravity" explains nothing in terms of why things fall. It is just a label – we might as well yell "INFINITY"! Our understanding of nature's laws is still the same.

In physics, how we define or label a concept often matters far less than how we *measure* it. *Experiment provides the answer!* It is not possible to talk one's way around nature's laws for long, though many have tried and continue to peddle their nonsense even today. All our labels and definitions are useless unless repeated testing provides supporting evidence for using a particular definition. For now, let's focus on the three-dimensional space we live in, and explore how we might usefully measure it. It could be that there are many more dimensions of space, somehow folded in on themselves so that we cannot see them but exist nonetheless. There are many physicists today studying the notion of a universe with as many as 11 dimensions, as well as the thought that there might be multiple *universes*. We start with something like space, and we think it is easy to think about because we live in it throughout our lives. But what do we really know about space, and what are we (perhaps) missing in a full understanding of it?

Evidence indicates that there is no point in space that is somehow special in relation to all the others, a condition which is often codified by saying "space is isotropic". That means we can choose the origin of our 3-D world to be at any arbitrary point that suits our fancy, and often that choice, if done wisely, makes solving a problem much easier. Every point in space can be ascertained by identifying the three coordinates, x, y, and z, in relation to the origin we choose. There is another way to establish a point in space, using the idea of a sphere. Imagine a sheet of paper splitting a basketball exactly into halves or hemispheres. To locate any point on this basketball, we need to know three things: the radius of the sphere and two other angular measurements, all measured from the center of the sphere, which of course is the origin. Rather than do the work for you, I leave it to the reader to figure out how to visualize and define those two angles. If you get stuck, find a basketball, remember the coconut trap, and have at it. If all else fails, look up polar and spherical coordinates!

For hundreds of years people have imagined a world in which there are only two spatial dimensions – a region called flatland. In this world, there is only length and width, no height. The concept of a sphere is difficult to conceive in such a place; in fact, were a sphere to sail up into and through flatland, think of what a fantastic scene that would make! First, you would see a point suddenly appear in your flatland, and then a circle would appear that keeps growing in diameter for a while, until the circle starts to shrink in diameter until it gradually just makes a point in your world, then disappears. It is fun to think of a place like that, but it wouldn't be much fun living there. If we call moving to the right along this surface "east" and give it the label "x" for horizontal motion, we can also do the same for vertical motion up along the surface and call that "y" or "north". The direction "west" is then just given by negative x, and "south" by negative y.

For centuries, many people believed we live on a flat surface, although most historians suggest that this belief was not as widely accepted as has often been portrayed. There are many ways to prove that the Earth is not flat: the shadow of the Earth passing across the Moon during a lunar eclipse is curved like a circle, and the mast of a tall ship at sea disappears gradually over the horizon, not all at once. But here is one approach you may not have come across, in the form of a puzzle. At what point on the Earth's surface can a person walk three miles due south, then three miles due east, then three miles due north and end up *exactly* at the starting point? Can you see this would be impossible on a flat surface? Yet there are an *infinite* set of points where one can do this on our planet! Now solve the puzzle, and realize in the process we live on a curved surface.

When we define a phenomenon we experience around us, we impose *our* reality on *it*. Therefore, the "trick" is to define space in a way that can be used, and then ensure that everyone will operationalize that definition in exactly the same way. This idea of establishing a standard unit for measuring space took centuries to develop, and it still causes some confusion today when people mix up their units because they get caught between different systems of measurement. We need to chop off a chunk of space, give it a name, develop a process so that everyone can reproduce this finite chunk of space, and then use it to measure distances in any one of the three dimensions we live in. Once we have a standard, *limited* distance, we can use it to discuss *any* expanse of space. If you think this is not so hard, imagine trying to convey what "one foot" means to a being living on another planet far from Earth. How would you do it? If you think this is not important, imagine trying to build just about anything without a standard unit of measure for the space it will occupy. Think about bridges, homes, computers, airplanes, clothes, shoes, hats, and nearly everything else in our world being created with no standard unit to measure space. Most of these things would fail, not work, or cause incredible chaos in such a world.

In many ways, it is a beautiful thing that there are so many languages spoken on Earth, along with a vast array of cultures and customs. However, when it comes to defining a standard for measuring space, it will be a tremendous advance in civilization when we develop one standard that is used by everyone. We are not at that point yet, though the scientific community is leading the charge! The goal of a standard unit for measuring space has a long and colorful history. We have encountered the cubit, yard, foot, fathom, league, rod, line...one gets the picture of dozens of words, each with its own standard representation of space. One might also picture the chaos that ensues when we try to communicate various standards for measuring space across different languages and cultures. It's a colossal waste of time and energy, and some of it persists even in more "advanced" technological places like the United States. Yet around 1792, as a new republic was struggling to establish itself here in North America, France was emerging from its own bloody revolutionary war. Change was occurring in societies all over the world at this time, and one of those changes, originating in France, was to supplant the many conflicting and confusing measuring systems with a single, superior one: the metric system. In a stroke of brilliance and as a nod to keeping the math simple, this system was based on powers of ten. The standard unit was called the meter, hence the centimeter is 1/100 of a meter, the kilometer 1000 meters, and so on. Because of the ease in which conversions can be done in this system, it evolved into the "System International" or SI units which are used throughout the world by scientists to this day, despite the reluctance of a few other fields (and cultures) to relinquish their attachments to the foot, pound, slug, yard, miles, and so forth.

Before we delve into the evolution of the standard unit of space, the meter, it is astounding to note that even though we cannot physically measure the largest or the smallest chunks of space with a standard meter stick, we *can* conceptualize them and we *can* infer their magnitudes through some very clever and interesting physics. Therefore, to appreciate just how extraordinary this feat of measurement is, we will take a journey through the largest chunk of space there is - the universe itself!

A Trip through the Universe at the Ultimate Speed

Author's note: I urge you to have a small notebook or pad of paper (and a pen) handy as you read. You will encounter puzzles and have many questions: writing these questions down is extremely important if you want to get the most out of our journey and this book.

Suppose you own a small rocket with an infinite supply of fuel. First, you might want to patent the rocket and adapt it to manufacture automobiles so you can live the rest of your life in leisure! Maybe you are more curious than practical though, and instead you decide to embark on a mission traveling through the universe. Interestingly enough, there are some "road maps" you can follow, and the plan is to take a route that travels past as much fascinating stuff as possible.

So, you hop in your rocket, loaded with food, water, and all the necessary provisions, and off you go. The rocket will start from rest, of course, and we will allow it to speed up gradually so you don't get whiplash. The accelerator is adjusted so that every second you gain 10 meters per second of speed. This rate of change in your speed is about what you are feeling now as you sit in the Earth's gravitational field, so you will be comfortable in your journey, since you are pressed into your seat with as much force as your own weight. Every attempt will be made to keep the rocket traveling in a straight line, because you know when you zip around a corner in a car you certainly feel additional effects that you might not care for too much! Also, you may temporarily feel forces acting on you that add up to more than your own weight, especially when you fly by a massive star. Since we know of no way to shield you from the effects of gravity, we'll do our best to make sure you survive those brief encounters. This trip isn't as easy as it may seem at first glance.

With the accelerator stuck on gaining 10 meters per second of speed every second, you find that after the first day you are traveling at around 864,000 meters per second. That's about 540 miles per second, or just under 2 million miles per hour. Hey, not bad! The fastest vehicle that we have traveled in as humans goes around 8 miles per second, so already you've broken a bunch of world records. Forget about the speed of sound: at room temperature sound travels at a pathetic two-tenths of a mile per second, or around 760 miles per hour. Your rocket, if it was sailing around here on Earth, would have broken the sound barrier only 35 seconds into the trip! We travel on, keeping the rocket hurtling through space in a straight line, with all the comforts of home and in complete safety. As long as we don't hit any little rocks, that is. If a pebble hits your windshield while you are driving down the highway at 60 miles per hour, it can easily crack the glass. Imagine what a pebble could do if it hit your rocket while you are cruising at 2 million miles per hour! There would be nothing left but random atoms floating through space. And we are just getting started. So, our rocket needs to have a space dust shield to prevent any such disasters. Of course, we have one of those – state of the art, too.

Each day we pick up the same amount of speed, and therefore after day two in our journey we are moving at 540 times 2, or 1080 miles per second. Obviously, we are covering some big distances at this point, but more on that later. Because around day 35 into our trip, something extraordinary and deeply puzzling happens. Actually, observers on Earth have been monitoring this effect for some time, but now the data is incontrovertible. We can't seem to speed up any more, according to their instruments. Our accelerator looks fine: it's pointing at 10 meters per second gained every second, but we are not gaining speed now! We have hit a barrier and we can't go any faster. That's just the beginning of this bizarre turn of events. Radio transmissions from Earth are also telling us that they are monitoring our clocks on board and our clocks are running substantially slower than identical clocks sitting there on Earth. We will leave these puzzles for now because we want to focus on the journey at the moment, but we will return to all of this and more, down the road. The bottom line is that we cannot travel any faster than our present speed, the speed limit of the universe: 186,000 miles per second.

Probably you know that the speed limit of the universe is the speed of light. It appears for the moment that *nothing*, whether it be a rocket, an atom, or a communication signal, can travel faster than the speed of light. (The reader might want to do a little research on something called "quantum entanglement", however, just for kicks.) Later on in the book, we will investigate how we first measured this incredible speed, and who was the first person to come up with an actual number for the speed of light. For now, concentrate on the remarkable magnitude of this speed: 186,000 miles traveled every *second*! If you could move around the Earth's surface at the equator in a circular orbit at the speed of light, you could orbit the Earth over *seven* times in one second! Some physicists think there may be particles, called tachyons, which can only travel faster than the speed of light, so that the speed of light cleaves the universe into two "worlds" that can never know one another: one faster than the speed of light and the other slower. We have no proof at this writing, however, that tachyons really exist, so for now this field remains one of conjecture and theoretical speculation only. Your rocket is therefore stuck at "c", the symbol we use for the speed of light. Actually, your speed is a tiny bit less than "c", and can never reach the speed of light no matter what you do. Make a mental note of that somewhere in your log book, because we will come back to that idea later on as well.

The space shuttle travels just under 5 miles per second while in orbit, depending upon its mission and orbital altitude. Your rocket is traveling over 37,000 times faster than the shuttle! We are now ready to start on our journey through the universe, and we will go at the speed limit of the universe because there is a lot to see and we don't have an infinite amount of time on our hands. We swing our rocket around and start the journey at our Sun, traveling at 186,000 miles per second. The Sun is not stationary: it rotates on its axis once every 25 days, and it travels around the center of the Milky Way galaxy in a roughly circular orbit (dragging all the planets including Earth along with it) once every 230 million years or so. Each *second* the Sun releases more energy than has been used by the human species for our entire history (a history which is a very tiny slice of time in the scheme of the things).

Our Sun has been around for billions of years, and we think after a few billion more years it will blossom out into a red giant and swallow the Earth in the process. But don't worry, you still have time to go to college, or walk along a beach, or live many lifetimes. And perhaps in a billion years our species will inhabit other solar systems with younger stars that have a long future ahead of them as our Sun does now. There are many different fates possible for stars; the mass of the star is the primary factor that determines which fate the star will "select". Our Sun is called a main sequence star: average in mass and with a surface temperature that produces a lot of yellow and orange light. The core of our Sun is so hot that if we could make an indestructible pebble as hot as the core, we could fry hamburgers with that pebble even if it were as far as 100 miles away. We'd fry, too, but don't worry about the details!

If the Sun were about the size of a large room, then Jupiter would be about the size of a basketball, the Earth would be the size of a grape, and Pluto would be like the seed within the grape. But it's the energy of the Sun that most boggles the human mind! Every second, 4.5 million tons of matter fuses into energy, streaming from the Sun in all directions. We didn't understand the source of the Sun's energy until midway through the 20th century. Before that time, we thought its energy might come from some kind of combustion of chemicals, like burning logs in a fireplace, but those calculations showed that chemical combustion could never account for the vast amount of energy being supplied by the Sun for billions of years. Now we know the process of nuclear fusion is responsible for the energy output of all stars, and in a later section we will look deeply into this process, because if we can figure out how to harness fusion here on Earth, our lives and our civilization will be forever changed and improved. And so we blast by the Sun at the speed of light, heading outward through our solar system.

One of the interesting things about our solar system is that all the planets "live" in the same orbital plane. Put another way, if we placed an orange on a large table and think of the orange as our Sun, then all the planets would exist on the surface of the table as well. No planet would be beneath or above the surface of the table. This gives us a clue as to how the planets may have formed: possibly as gravity pulled the pieces of the Sun together to form a star, the

Sun began to spin faster and faster as figure skaters do when they draw their arms inward. Perhaps some chunks of matter flew off this spinning disk to later become planets – this would explain why all the planets lie in roughly the same plane. A little over three minutes into our flight from the Sun at the speed of light, we encounter the planet Mercury. (Remember we are taking a route that will afford us the most interesting sights along the way. While the planets rarely "line up" so we can go in a straight line, let's assume for simplicity we can navigate our way through the solar system flawlessly!)

At the speed of 186,000 miles per second, we travel from the Sun to its nearest planet, Mercury, in about 3.3 minutes. This works out to about 37 million miles. Mercury has no atmosphere and no moons; it has a metallic core that is 5.4 times denser than water. The temperature range on Mercury is extreme: from 660 degrees to minus 270 degrees Fahrenheit. There are cliffs and craters almost 800 miles across, and we think the planet was formed about the same time the Sun was born – roughly 4.5 billion years ago. Mercury takes 88 days to circle the Sun once, and it rotates once on its axis every 59 days – so its day is nearly two-thirds as long as its year! A person weighing 200 pounds on Earth would weigh about 80 pounds on Mercury's surface. There is no life (as we know it) on Mercury.

Continuing in our travels at "c", we pass the planet Venus at a little less than six minutes into our flight. Venus has the hottest atmosphere of all the planets due to the gasses held within its gravitational field: energy from the Sun is trapped by these gasses in a colossal Greenhouse Effect that raises the temperature of Venus to as much as 900 degrees Fahrenheit. Venus is the brightest planet as seen from the Earth, sometimes erroneously referred to as the "morning and evening star". Planets do not undergo nuclear fusion; hence they may reflect and absorb radiant energy but they do not produce such energy through the fusion process, and they wander through the nighttime sky (which is how the word "planet" got its name). So, planets are not stars, and falling "stars" are chunks of rock (meteorites) hitting the Earth's atmosphere at breakneck speeds and burning up in a streak of glory. Venus has no moons and a metallic core, like Mercury. The really wild thing about Venus is that its day is longer than its year! It takes Venus 225 days to orbit the Sun once; but it takes 243 days for it to rotate once on its own axis. Between one Sunrise and the next, more than a full "Venusian" year passes by. Imagine going to the beach on Venus as the Sun stays out for over half a year while you sit on your blanket and roast.

We travel onward and exactly 8 minutes, 20 seconds into our flight, we pass by the Earth. Zipping along at 186,000 miles per second, it takes light from the Sun over 8 minutes to reach the Earth. And that's a pretty amazing thing if you think about it. Because it means the Sun might not exist *now*. In fact, we never see the Sun as it exists at exactly this moment in time. We see it as it was 8 minutes ago; for all we know, the Sun just blew up and it will be 8 minutes before we all get obliterated. That could be a strong argument for living in the moment! The mystery is much deeper than that, though, for you have to understand that we never see *anything* "now". Look at the back of your hand: it takes time for the light bouncing off your hand to reach your retina and be processed by the brain. Not much time to be certain, but it still takes time for the light to reach your eyes. So, you don't see the back of your hand "now", ever, and you never will! The farther away an object is, the further back in time we see it. The Sun is 8 minutes back in time from Earthlings. A star that is far enough away so that it takes 158 years for its light to reach the Earth may have exploded in a supernova long ago, we just don't know it yet. And if there are planets around that star, with people and telescopes, they are seeing the light that left the Earth 158 years ago…which would be the year 1862. That means they could be eyewitnesses to the Civil War – watching the Gettysburg Address as you read this! It's all very strange – and wonderful at the same time. By the way, once we've traveled through the entire universe as we know it, I think you'll be convinced there is definitely life beyond our solar system: the really hard part is finding it.

We are passing the Earth and there is much to consider. If we were 1% closer to the Sun or 1% farther away, there would be no life on Earth. Look up at the sky – look out your window – look at the pavement or your friend or whatever: it's a miracle we're here, at exactly the right place and time, to witness it all. The Earth is traveling around the Sun at 66,000 miles per hour, and recall that the Sun is moving through the galaxy as well. You may not be moving along a highway right now, but you are attached to the Earth and the Earth is moving. Nothing is stationary in this universe; there is no frame of reference that is absolutely motionless relative to all other frames. Everything moves. *Everything*.

Two more seconds into our flight and we fly by the Earth's moon. At 186,000 miles per second, it doesn't take long to get from the Earth to the moon! On the moon, we notice human footprints that will never erode or disappear unless something crashes into them from outer space: there is no atmosphere, no water, no wind. While there are theories as to the origin of the moon, many scientists still classify the formation of the moon as an ongoing puzzle.

At 12.5 minutes into our journey from the Sun, we observe the planet Mars. Our odometer reads 140 million miles traveled in this short period of time. Mars has volcanoes reaching 17 miles high – roughly three times as high as Mount Everest. It has two moons and canyons 3,000 miles long. Mars has ice caps composed of water and carbon dioxide. Our unmanned probes to Mars continue to uncover many startling aspects of the Martian surface, and the possibility of some form of life on Mars remains open for debate. You may have seen in the movies how there is a time delay sending signals from the Earth to spacecraft orbiting Mars.

Hollywood often botches the science in favor of what they might consider as more fun (to me the fun is in the actual science), but they have this one right. Now imagine we build this contraption and send it from Earth to Mars without any humans on board. This is very difficult because Mars is moving around the Sun in a different orbit than we are, so it's a little bit like trying to throw a baseball across to your pal on the other side of a spinning merry-go-round, using radio signals to guide its path. And the radio signals take time to reach the craft so you have to take that into account while everything is in motion. Then you need to achieve orbit around Mars, and then find a way to land this contraption so we can gather soil samples and the like, all by delayed radio signal. Imagine doing that! You don't have to – we just did it. Using science.

As we fly outward from the Sun beyond Mars, there is a wide gap in space and time before we reach the next planet. This gap is filled with asteroids and, as was observed previously, any one of those asteroids would send us into oblivion should we be unlucky enough to suffer a collision while traveling at 186,000 miles per second. The largest planet in our solar system, Jupiter, looms on our horizon about 43 minutes into the journey. Jupiter has 16 moons plus rings, and a 200-pound person would weigh almost 600 pounds on its surface, if we can even define a surface for this gaseous giant. It takes Jupiter 12 years to orbit the Sun once, and it rotates on its axis once every 10 hours. Galileo, one of the giants in the history of physics, advanced our knowledge about the moons of Jupiter significantly. The infamous "Red Spot" that migrates around the equatorial region of Jupiter is actually a perennial hurricane of sorts – the "spot" itself is bigger than the Earth.

Saturn, with its fantastic rings numbering over a thousand, flies by our window around 79 minutes into the flight. Saturn is less dense than water, so if we could find a big enough pool, the planet would bob up and down in the water like a cork. Saturn has at least 23 moons, and it takes

29.5 years for it to orbit the Sun once. We are currently studying the rings of Saturn to determine their origin and composition. Uranus is in front of our view window 160 minutes into the flight, still pressing the accelerator to the floor and (helplessly) obeying the speed limit of the universe. This planet is barely visible to the naked eye and wasn't discovered until 1781. The striking thing about Uranus is that it lies on its side as it orbits the Sun, almost as if something knocked it off its feet and made its rotational axis align itself parallel to the plane of orbit rather than approximately perpendicular like all the other planets. We don't know how this happened.

The last two planets that we know about, and there could easily be ten or more planets in our solar system because they are too distant to see even with today's powerful telescopes (remember that planets do not give off their own light), are Neptune and Pluto. Although Pluto is usually the farthest planet from the Sun, because of its very eccentric orbit it sometimes curves inside Neptune's path – this happened just recently for a time. Having an eccentric orbit means that the path traced by the planet around the Sun is not circular; rather it is oval or elliptical in nature, like an egg. More often than not, we would find Neptune about 240 minutes into our trip, and finally Pluto 330 minutes traveling at "c". Pluto has a moon, Charon, that is one-third the size of Pluto, and a 200-pound person would weigh 6 pounds on Pluto's surface. It takes Pluto 248 years to orbit the Sun once.

I haven't said anything about this until now for the sake of clarity, but a couple of items need to be addressed. First, when we say it takes 248 years to orbit the Sun once for Pluto, we obviously mean 248 *Earth* years. Plutonians (now there's a word!) would call one trip around the Sun one year, just like we do, but their year is 248 times longer than ours! Second, all the times and distances we are recording are based upon what an observer sitting on Earth would "see"; it turns out that clocks on our rocket do not tick out time the same way as clocks that are not moving relative to the observer. So, the time it takes to do this little trip of ours is relative to the observer! If that confuses you, join the club. We will dive into this relativity stuff later on.

As far as we know at this writing, Pluto is the last planet we would see before passing out of the solar system. Traveling at 186,000 miles per second for 330 minutes, we have traversed the entire solar system in less than six hours. The total distance we have covered is now about 3.7 billion miles. With great confidence, we forge ahead into interstellar space, searching for the nearest star to our Sun, Alpha Centauri. Given that it took only six hours to reach the outermost planet from the Sun, one would think it shouldn't take too long to reach the closest star and solar system to us. Think again. Alpha Centauri pops into our screen 4.3 *years* into the journey! Pluto was just a little baby step, as it turns out. Hang on as we blast through the Milky Way galaxy at the ultimate speed.

Our solar system is on the outer reaches of the galaxy, and as mentioned previously it orbits the center of the Milky Way once about every 230 million years. Traveling at "c", it will take us 25,000 years to reach the center of our galaxy – you begin to see why we need an infinite supply of fuel (and patience). Remember, though, that the 25,000 years is what an Earth observer measures – those of us on the trip in the rocket would measure a far different (and shorter) time to reach the galactic core. To pass out of our galaxy altogether on the far side from us would take about 100,000 years, as measured from Earth. The nearest large galaxy in our local cluster of galaxies is Andromeda (we have beautiful slides of Andromeda taken from powerful telescopes), and Andromeda is 2 million years away from us traveling at the speed of light, and currently on a collision course with our Milky Way galaxy. Each galaxy contains billions of stars, some like our Sun and some not, but nearly all have the potential of planetary systems orbiting around them. There are currently *billions* of galaxies, some like the Milky Way and some not, that we have catalogued at this writing. One begins to understand how the probability of life existing elsewhere hovers close to 100%.

What happens if we keep traveling onward, in a straight line, at the ultimate speed of the universe? Of course, we don't really know because we haven't done it, but we have some educated guesses based on a lot of data, logic, and consistent reasoning. We think the universe is somewhere between 13 and 14 billion years old. Recall that we never see anything "now" – it takes light some time to travel through any amount of space, small or large. The Hubble telescope can pick up light from distant objects in the universe with incredible sensitivity, since the Earth's atmosphere does not impair the telescope's view from its orbital perch above the Earth. Objects as distant as 13 billion years away from us have been observed through these telescopes, and you know what that means. It means we are literally seeing the beginning of time, because the light that left these most distant objects had to start on its journey 13 billion years ago to reach us now, and 13 billion years is just about how long we think the universe has been around.

It gets better: some physicists surmise that if we kept going at "c" in a straight line, we would eventually end up where we started! The universe sort of curves in on itself in four dimensions: the three spatial dimensions of height, width, and depth, and the fourth dimension of time. In an analogous way (but with fewer dimensions to worry about), if we walked around the Earth in a straight line without stopping, we'd come back to where we started in one big circle. While it might be fun to wonder what there is "outside" the universe or beyond its "edge", the question has no meaning. The universe does not exist in some static, giant cavern of empty space. All the objects of the universe, and all the space in between, exist within the universe. And we know the universe is expanding still, at a remarkable rate. After all is said and done, we have traveled at the ultimate speed of 186,000 miles per second for over 13 billion years…past planets and exploding stars and clouds of interstellar dust becoming new stars…only to arrive back at the place we started. Maybe, that is. The certainty I can offer you is a little different: the mystery of our existence in this universe is the greatest of all mysteries. I don't believe one can live a full life without contemplating this mystery once in a while, because that kind of contemplation yields a very deep and lasting appreciation for all of life – yet another great circle that encompasses all we hold dear.

After we return from our journey, the problem before us is to define the standard unit of measure for space that is accessible to everyone and reproducible by following a straightforward set of rules. The French Academy of Sciences was comprised of many scientists with probably hundreds of possibilities for this standard unit, but over time two major candidates emerged. The first of these was to use the pendulum. About the year 1800, scientists were quite familiar with the physics of a simple pendulum, but we weren't sophisticated in our measurement of time. That was a major stumbling block, since the length of the simple pendulum along with the value for the acceleration of gravity at a given latitude is what determines how long it takes for the pendulum bob to swing through one complete cycle. Working backwards, if we know the time it takes for one cycle accurately, we can then deduce the length of the pendulum with some precision assuming we have ascertained the acceleration of gravity through freefall experiments. Yet measuring a time interval precisely is not as simple as it might seem. A century before this Galileo was using the steady drip of water or his pulse to measure the time it took for objects to move down an inclined plane. One can see the difficulty in using such methods, especially if we are using our pulse to measure something and get increasingly excited while doing so!

The second candidate set forth as the basis for the standard meter was to use a section of the Earth, and this was the idea that was pursued. Just prior to 1800, the process of measuring and locating exact points on the Earth's surface, along with the distance between those points, was well established. As a matter of historical interest and as an example of sheer genius, consider the fact that in the third century B.C. a Greek mathematician and astronomer named Eratosthenes discovered an experimental method to determine the radius of the Earth, hence the size or circumference of the Earth, using sunlight, shadows, and a long walk at a particular time of year. Unfortunately, the exact methods he used to establish the Earth's radius were lost in the fog of ancient history, but simplified versions of his work survived. Of course, this implies that the Greeks knew the Earth was spherical, not flat, and it is a brilliant piece of analysis that yielded a value for the Earth's radius that is quite close to the one we use today. (Though the Earth is actually shaped like an oblate spheroid, we can employ the equatorial radius, and that value is 6.38×10^6 meters.)

Since the process for measuring sections of the Earth was well known by 1800, it remained only to find a suitable length that was based on the size of the Earth. The decision was made to call the distance from the equator to the North Pole ten million meters, as this would cause the length of one meter to be easily manipulated by people wanting to measure ordinary sized objects and expanses of space. A small arc of meridian was measured from Dunkirk to Barcelona between the years 1792 to 1798, with the flattening of the Earth taken into account as one approaches the poles. With an excellent value for the length of one arc of meridian, it was easy to extrapolate to ninety degrees of meridian, take one ten-millionth of that amount of space, and call it one meter! This length turned out to be a bit more than the common English "yard", and is in fact 39.37 inches. If you ever are sitting at the equator, imagine laying down 10 million meter sticks, end to end, pointing due north. Walk along those meter sticks (a feat destined for the record books) and eventually you will be at the North Pole!

The French had now established a standard unit of measure for space, the meter, that was seemingly indestructible, available to everyone at least in principle, and reproducible by everyone if one had the necessary resources. One can readily see that it is impractical to expect people to be out there measuring out an arc of meridian every time the standard meter is required, so it was soon decided to manufacture a platinum bar to represent the meter. Such a bar was made, and copies distributed gradually throughout the world. The original platinum bar was placed in the custody of the Archives de France, and was stored in Sevres, a town near Paris. This standard meter was used for over 80 years, and even though it was later discovered that the platinum bar was too short by a fraction of a millimeter with respect to the equator to North Pole distance, the error at the time was of little consequence. Such an error today would be catastrophic in terms of the operation of high precision machinery.

Small changes were made along the way to this metal bar, such as improving its mechanical properties by making the metal bar 90% platinum and 10% iridium, thereby reducing the effects of expansion and contraction due to temperature changes. Fine lines were engraved on polished areas of the bar to mark off one meter, and the uncertainty evolved to a remarkable one part in ten million. That means if one were to lay down ten million standard meter sticks from the equator to the North Pole, one would be a *maximum* of one meter away from the actual pole. Remarkable progress to be sure, but issues remained. This platinum-iridium bar could be destroyed by fire or war, or it could be stolen. Making copies from any standard metal bar necessarily introduces error, as does trying to duplicate the measurement of one degree of arc in a meridian. When new parts machined to a given specification don't fit precisely into a machine, everything comes to a grinding halt. This notion of a standard meter is not some abstract, fanciful notion that resides in the exclusive province of scientists: it is the heart and soul of an advanced, industrial and technological civilization. Still, this standard for a chunk of space was about to undergo even more changes, each of which would enhance the precision of the standard meter.

Some of the most rapid and revolutionary growth of scientific knowledge in human history has occurred in the past 150 years, so perhaps it was inevitable that the meter would reflect this evolution along with everything else. With the discovery of atomic spectra and the associated emission lines of various elements, a virtually indestructible atomic standard for the meter was established: 1,650,763.73 wavelengths of the radiation corresponding to the transition $2p_{10}$ to $5d_5$ in an atom of krypton-86. Essentially this means when an electron in an excited state jumps down to a lower, more stable energy level in the krypton atom, a photon of a very specific frequency and wavelength is emitted from the atom to preserve the conservation of energy principle (something we will explore toward the end of the book). The wavelength of this photon is precise and measureable (and very small), so over 1.6 million of those wavelengths will comprise a meter. This improved the precision of the meter by a factor of ten, and the standard was clearly invincible in the face of war, theft, fire, or even a shift in the shape of the Earth.

This might seem like the end of the story regarding defining space in a way that can be measured repeatedly with an agreed upon standard, never losing sight of the fact that the entire metric system is based on powers of ten, making it significantly easier to use than previous awkward systems (12 inches to a foot, 3 feet to a yard, and so forth). One more iteration remained, however, and it arrived in October of 1983: one meter is now defined as the length of the path traveled by laser light in a vacuum during a time interval of 1/299,792,458 seconds. Be on your toes dear reader, because this current standard for the meter clearly leads to a multitude of new questions and challenges. How does one measure the speed of light to that degree of precision? What is meant by the time interval "one second"? That is a perfect segue into our second major concept in physics: time. Before venturing out on that journey, however, there are a few items remaining regarding this thing we call "space" that need addressing, and one more voyage into space of a far different nature than the observable universe. At the moment, it is hoped that one can comprehend how ideas about space are inextricably intertwined with our thoughts about time, to the point where we now define space using the measuring of time. And the speed of light: why does that keep rising to the surface in just about every physics concept? One of my earliest memories of teaching involves a student, James Q, who jokingly would blurt out "the speed of light" whenever the class was stumped about a question I had asked. His "theory" was that the speed of light had to be the answer in most cases! We had a lot of fun with that, and looking back, I think James was onto something.

We have seen that space is three dimensional, so that any point in space can be defined by a set of three numbers which we call coordinates (x, y, z). Some textbooks will use (i, j, k) notation; others might use spherical coordinates that were mentioned previously. But they all do the same thing: pinpoint a location in space. When we measure the space between two established points in space, we call that "distance" and use the meter as the agreed upon standard. This is further defined as a "scalar" quantity, meaning we only care about its magnitude – how big or small it is – and no other information is required. Therefore, if I am driving to Boston, I say the distance traveled is 54 miles, which I can then convert to meters by knowing that 1 inch is 2.54 centimeters and then doing the necessary dimensional analysis to arrive at the number of meters. Yet if I want to be more specific, and actually *get* to Boston, I need to specify my direction as well as the distance traveled. When direction is specified along with the distance between two points, that is defined as displacement. Further, any physical quantity that has direction associated with it is called a vector. Obviously, if someone is lost in the woods or out at sea, I want to know the correct *displacement vector* to rescue them, not just the distance they are away from my present position.

Now an intriguing question: is space absolute? By that we mean, does everyone agree that the space called one meter will be the same to all observers, regardless of relative motion between those frames of reference? Miraculously, astoundingly, the answer is no. It turns out that space contracts in the direction of motion, so that if one were to somehow measure the length of a meter stick that is traveling toward us at a high rate of speed, the meter stick would be shorter than its "at rest" length. This is not an optical illusion; it is very real, and this phenomenon has been verified experimentally many millions of times over the past one hundred years, a span of time when we actually had the technology to measure this effect. I will offer you one experimental verification of this length contraction. There is a particle created in the upper atmosphere when high energy cosmic rays emitted from the Sun interact with particles within the atmosphere: the muon. We know from lab experiments with these muons that they are very unstable, and when at rest in the lab, decay into other particles very quickly (2.2 microseconds). Based on this knowledge and knowing where in the atmosphere the muons are created, we would expect very few to reach the Earth's surface before decaying. We run the experiment, always reminding ourselves that nature does not care about our beliefs – she makes the rules. It turns out that far more muons reach the surface than we would expect if space (and time) are absolute. The muon "sees" a space contracted distance to travel to the surface, or alternately, lives longer before decaying than what it did in the lab while at rest. Either way, length contraction or time dilation are the only way to explain the experimental results of so many muons reaching the surface. The facts are in: relative motion alters our very notions *and measurement* of space and time, and all of it is hinged on the fact that the only absolute there is…you might have guessed…is the speed of light! The closer one moves to the speed of light, the more pronounced the effects become.

This is truly bizarre, and yet why should nature conform to our *belief* that space and time are absolute? It appears the fabric of space-time is affected by motion through it; we do not ordinarily notice this because the only metric nature cares about is a huge number: the speed of light. There is a great book I am thinking of titled *Mr. Tompkins in Wonderland* by George Gamow, in which we live in a wild world where the speed of light is quite small, maybe 10 mph or so. In this world, everyone knows about time dilation and length contraction because it is observable all the time and is just "common sense". But we don't have to travel to this wonderland to make note of an everyday occurrence that rests on these experimental results: your beloved GPS system!

To locate your position precisely and navigate you through the world, satellites zipping in orbit around the Earth triangulate (actually four satellites are used) your coordinates and thus locate where you are with great precision. Without taking into account length contraction and time dilation, along with the effect gravity has on the measurement of time, your GPS system would be totally useless – missing your location by many miles. Extremely precise atomic clocks on these satellites are ticking out time differently than identical atomic clocks on Earth, and we must account for all of this if GPS is to work. We begin to see that all those shows about supernatural beings and mysterious powers are far less interesting than what we actually experience on a daily basis. As the saying goes: "truth is stranger than fiction". I would add, reality is miles ahead of all the purported "strangeness" dreamed up by humans, no matter the reference frame. Tolkien's *Lord of the Rings* is an absolute masterpiece, and I loved every word in it; I just find myself wishing at times that we shared the same fascination with nature herself, whether its muons or contemplating our existence amongst billions of galaxies!

There is another peculiarity regarding the concept of space that no one has figured out yet. The description of this puzzle starts with hydrogen gas contained in a sealed glass tube. Place this tube under high voltage, and the hydrogen inside glows in the violet and blue part of the spectrum, predominantly. Upon closer inspection, with an instrument called a spectrometer, when the light from this excited hydrogen passes through a narrow slit on the spectrometer, four distinct lines of light appear. This is called an emission spectrum, and the lines are formed when the lone electron in the hydrogen atom "jumps" from the higher energy orbital levels it attained because of the high voltage, to lower (more stable) energy levels. Specifically, this is the Balmer Series, which has been known and observed for over a hundred years, but was finally (partially) explained and predicted by Niels Bohr in the early 20th century. The energy levels of the electrons in any element are quantized, meaning these electrons can only assume specific energies. This is a very strange situation: imagine if the speed of an automobile was quantized in "lumps" of 5 mph, so we could only go 5, 10, 15, 20 and so on mph and nothing in between!

Therefore, when the single electron present in the hydrogen atom "jumps" from energy level 5 down to level 2, a precise amount of energy difference occurs (literally subtract the difference in energy levels), and that precise energy difference is carried away from the atom in the form of a photon of light. Since this photon has a specific energy, it also must have a specific frequency and wavelength, and in this case, that corresponds to a violet emission line that is visible in the spectrum and is 434 nanometers. The other lines in the Balmer series correspond to energy transitions between energy level "n" to 2, where n is integers from 3 and higher. As one more example, the transition from energy level 3 down to 2 produces a photon of wavelength 656 nanometers, which we see as a bright red line in the spectrum. Energy transitions from 7 or higher down to 2 produce photons in the ultraviolet region, which we cannot see but are nonetheless there (and should not be looked at without protective glasses!).

Each of the over 100 elements known to us in the universe has a distinct, unique spectrum that serves as a blueprint for the composition of the material. This is what makes spectral analysis so valuable in forensics, astronomy, or *any* study of the composition of matter. Just as the uniqueness of density can be used to identify solid, liquid, and gaseous materials (assumed more or less pure), we can use spectrometry to figure out what stars are made of or what chemicals are present in any substance. It's a wonderful diagnostic tool with thousands of applications. You might be wondering what in the world this has to do with space. Onward to the next part of this unsolved puzzle: the Doppler Effect.

When a car is approaching you on the highway, the pitch of the sounds you hear, whether from a blowing horn or the tires on the road, is higher than when the car is at rest, and then the pitch gets lower as the car recedes from you. Be careful: we are not discussing the intensity or loudness of the sound: clearly things get louder as the source of the sound approaches and softer as they recede. In this case, we are referring to the *pitch* or frequency of the sound waves heard. This shift in frequency due to relative motion between the source of the wave and the observer is an aspect of *all waves*, not just sound, and is dubbed the Doppler Effect in honor of the person who studied this phenomenon in the 19th century, Christian Doppler. The applications of this pattern in nature number in the thousands: Doppler radar weather, radar traps for speeding vehicles, monitoring fetal heart rate, monitoring blood flow in arteries, clocking the speed of a fastball, and so much more.

Now we are ready to get into the heart of this puzzle concerning space. In the 1920s and 1930s Edwin Hubble was trying to ascertain if clouds of interstellar dust (called nebulae but many are actually galaxies) were located in our Milky Way Galaxy. The thinking was that perhaps these clouds were in fact much more distant, beyond our own galaxy. By studying photographs of Andromeda Galaxy and stars with varying brightness, called Cepheid variables, he determined that at least one in particular was around 900,000 light years away. Recall from our trip through the universe that the entire diameter of our galaxy spans "only" 100,000 light years, so it was clear that the Andromeda Galaxy was much farther away from our Milky Way than previously thought. It turns out from more investigation that Andromeda is about 2 million light years from our galaxy, and there are billions of galaxies far beyond that, each with many billions of stars and planets. Moreover, and this is where the Doppler Effect and the hydrogen spectrum enter into the picture, it was discovered that many stars are composed of mainly hydrogen gas (by identifying the unique spectrum) and that the visible lines in this spectrum were shifted from the usual pattern seen in the lab to longer wavelengths (lower frequency, hence the term "red-shifted"). This means that these galaxies are sending light that is Doppler shifted to lower frequencies, which means the universe is expanding!

Think of a balloon with dots painted on it, and as you inflate the balloon the dots get farther and farther apart. Though this is not a perfect analogy, it best describes the expansion of the universe: it is not galaxies rushing headlong away from each other; rather, *it is the space between all the galaxies expanding rapidly!* One begins to see when you take this "simple" concept of space and dig deeply into what that means, the new horizons are endless. Now to the very core of the unsolved puzzle, because we have known the universe is expanding for a long time. What we didn't know until fairly recently, and what nobody has figured out since, is why and how the expansion of the universe is *accelerating*! The term "dark energy" has been conjured up to try to provide a conceptual model for what might be pulling the universe apart at ever increasing speeds, but we don't have a clue as to what "dark energy" is – yes, another example of a label being attached to something we do not understand!

Before taking a final trip into the smallest spaces we know, I want to relate a human side to this story, because I think it is important to recognize that *everything is connected,* including the people who work on such puzzles. Hubble did not work alone. His associate was a man named Milton Humason, an aspiring astronomer who began his lifelong passion with no formal schooling beyond his early teens. He became a janitor at Mount Wilson, the observatory where Hubble did much of his work, and quickly showed a profound interest and technical ability that earned him a staff position at the famous observatory. He went on to aid Hubble in much of his research and contributed greatly to all of the work being done in cosmology and astronomy, yet I am betting very few of you have heard his name mentioned. This is a story that is repeated many thousands of times in the history of human progress, regardless of the field of study. Far too many of these forgotten examples were women, people of color, or people not part of the mainstream of established "paths up the mountain". Even the great Isaac Newton, not always known for his humility, had the awareness to be quoted as saying "if I have seen farther than most, it is by standing on the shoulders of giants". In that sense, every person who works to understand our existence in deeper and more profound ways is a giant, and we are all connected in ways we most often do not comprehend.

This final portion of our investigation into the concept of space will be a trip into the smallest spaces we know. We need to develop a scaled model for this journey, so to begin we imagine a human being that is 25,000 miles tall. In this world, a piece of paper would be 2.5

miles thick, and the atoms making up the paper would be less than half an inch tall. The nuclei of these atoms would be close to the thickness of a human hair. If you wanted to know the thickness of a molecular layer, try this experiment: put a drop of olive oil in a large container of water and wait for it to spread out uniformly across the top of the water. We know the density of olive oil is about .916 grams per cubic centimeter, and we know the volume of a solid cylinder is $\pi R^2 h$ (R = radius of the top or bottom of the cylinder, h = height of the cylinder). Measure the radius of the circle of oil you see once it finally stops spreading on the water, and from that you can figure out the thickness "h" of the molecular layer. (Hint: you will need to measure the mass of this droplet, or find a way to measure the volume of the droplet. Ben Franklin did something similar to this, many moons ago!)

Is there anything smaller than the nucleus of an atom? Yes, since the nucleus is made up of protons and neutrons. Beyond that, we have evidence that protons and neutrons have internal structure and are comprised of three quarks each, though at this stage of our progress we have not observed an isolated quark or split a proton into smaller pieces. One can see that this process is analogous to walking toward a wall, with each space covered being half the distance as the previous one. In theory, we would never reach the wall this way, assuming we have no spatial dimension ourselves. There is a final step to take here, and it involves stars much larger than our own Sun, which eventually run out of nuclear fuel and then collapse on themselves due to the inward gravitational pull of their gigantic mass, since there is no longer any outward pressure being produced by the fusion process. This collapse continues indefinitely in some cases, creating those infamous "black holes" that populate the universe. Gravitational fields become so strong in these collapsed stars that light itself cannot escape, which is another puzzle since one might reasonably ask how something with no mass (light) is affected by gravity. More on that topic later. Questions abound when we investigate the physics of black holes. What is at the "bottom" of the hole, or does the question even make sense? The term "singularity" is used for the point in space where all the mass of the star apparently goes, which might serve as the smallest space of them all. Is that a rip in spacetime? Do our physical laws work inside black holes? Does the other side of a black hole spawn a different universe? We don't have the answers to any of these questions, but the evidence for the existence of black holes is overwhelming. Check out Cygnus X-1, a stellar mass black hole about 6000 light years from Earth, with about fifteen times the mass of our Sun. Every time we think we have "arrived" and solved all the mysteries of the universe, nature produces new evidence that our journey to understand all of its intricacies is still in its infancy. Onward to our second fundamental concept: time. We may see that separating "time" from "space" may not coincide with nature's preference in the matter.

TIME

The three-dimensional space we (apparently) live in requires three numbers to specify a location, and if a direction is specified it is termed a "displacement" and thus defined as a vector. Time is a scalar: it has no direction in the sense of up, down, left, or right, and it makes no sense to describe a time as 60 seconds at an angle of 40 degrees North of East. Time appears to move forward, since the only way we can make sense of it is to define its passage by marking events that are separated in time, in much the same way that we need objects to measure the space that exists between them. But "forward in time" is not a direction. We have come to understand that the laws of nature are inextricably tied to the laws, axioms, and methods of mathematics as described many pages ago. In the case of our concept of time, it appears the mathematical principles of probability rule the day, as they do in the world of the atom and in so many different aspects of our lives.

Drop a rock in a pond and watch the spherical wave-fronts ripple outwards in concentric circles of ever decreasing amplitude, until at some distant point away from the source, the ripples are imperceptible to even an astute observer. How wild would it be if someone sitting by a pond saw the process in reverse, and the ripples got larger and larger until the rock suddenly popped out of the water up onto land? I know many a golfer who would love to see this reversal in time! Why is it that autumn leaves never randomly, spontaneously assemble themselves into a nice, neat pile for us? What keeps all the air molecules in a room widely distributed so we can breathe easily, rather than at any given moment in time having them all "decide" to move to one side of the room? None of these processes are impossible, but the probability of their occurrence is so close to zero that the chance of such an observation would be about once during the entire age of the universe. Natural processes strive to achieve equilibrium, which is another way of saying "nature moves toward a state of maximum probability". This is the reason when one plays cards it is nearly impossible to get dealt the entire suit of diamonds in a game of "Hearts". There's only one way for that to happen in a scenario where millions of outcomes are possible. Understand that every microstate, that is, every hand of thirteen cards that is exactly specified by suit and number, is equally as probable as getting dealt the suit of diamonds. One can see, however, that there are millions of mixed hands possible, and that is why one nearly always sees such an outcome. Leaves scatter, sugar cubes dissolve and spread out, and highly ordered structures called "life" eventually die. Nature drives all processes toward equilibrium, maximum probability, and for that we coin the term "entropy". You will note that I have avoided the term "disorder" when discussing entropy, even though it is often cast in such language. This is because I do not know how to measure or operationalize such a vague notion as "disorder", *but I can measure probability.*

Ludwig Boltzmann knew this, and showed us how it all works for large numbers of molecules in a gas. On his tombstone is engraved his famous entropy equation: $S = k \cdot \ln W$. The symbol S is entropy, k is Boltzmann's constant, "ln" is a natural log to the base "e", and W is termed thermodynamic probability. Like all equations, this one tells a story: entropy is a measure of probability, and the most probable state is called equilibrium, and that is why nature tends toward equilibrium. There is a wonderful book titled *Five Equations that Changed the World* - entropy is one of those five equations. It is tragic that Ludwig took his own life in 1906, just moments before his statistical approach to physics found its wings and gained credibility. People were not anxious to leave the deterministic, cause and effect world of Isaac Newton and James

Clerk Maxwell, two giants of physics who we will meet down the road. Nature once again, however, dictated otherwise, caring not at all for what people were clinging to or how they thought the universe should operate. There are many arrows of time, all moving forward in a sense, but these are not vectors. Time is a scalar quantity, having magnitude only.

Once again, we approach a concept in physics by focusing on how we measure it. The unit for time has hinged on several phenomena in our history: dripping water, our pulse, pendulums, the seasons, and many more. Perhaps it was inevitable that we looked to the solar cycle as the first indestructible, dependable, and reproducible standard to measure time. We divided this sunrise to sunrise duration into hours, minutes, and finally seconds: 86,400 of them. The problem with this method of defining the unit of time as the second is that the rotation of the Earth is not constant, and in fact it is slowing down ever so slightly each day that passes. One can think of this much like a figure skater stretching her arms out while in a spin: there is a law of nature which we will explore later on that demands we spin slower as the mass distribution gets farther away from the axis of rotation. As the Earth rotates, and given that much of its surface is water, it develops a bulge at the equator and flattens at the poles, hence it must spin slower over time. This is a small problem, or maybe not a problem at all for most situations, but in our technological age that demands precision, it is a giant stumbling block.

For a time, we used the year 1956 as the average solar day so everyone was on the same page, but the standard eventually adopted as our technical knowhow progressed was to base the second on atoms. Specifically, one second is defined as 9,192,631,770 vibrations (periods or cycles) of the radiation corresponding to certain energy transitions in cesium-133. The precision of this standard is amazing, with an uncertainty of a few parts per one hundred billion. This means it takes over one hundred million years to be off by around one second. Obviously, these are expensive clocks, and we can depend on them (if we had one) to never be late for that dinner date! There are many such clocks now in the world, and the experiments we can do with them are superb examples of how advancing technology can catch up to scientific theory, which in turn generates new technology. Let us look at two of those experiments.

We have seen how space is not absolute – it contracts in the direction of motion, observers in different reference frames do not agree on the extent of space measured, and this effect becomes observable as one approaches the speed of light. It's reasonable to ask if the same is true about the concept of time. The first experiment involves loading a cesium atomic clock on a jet aircraft, flying it around the world a few times, and then comparing how much time elapsed on this clock versus an identical cesium clock that remained stationary throughout the experiment relative to the moving clock. The clocks are exactly synchronized at the start of the flight, and recall the remarkable degree of precision these clocks have. One might assume that these clocks would still agree, exactly, after the flight was completed, and that the amount of time should not depend on the motion of one of the clocks. The fact is, it does, and the clocks do not agree, and the moving clock ticks out time slower by precisely the amount as predicted by Einstein in his Special Theory of Relativity. (The General Theory of Relativity also enters the picture here, and we will discuss how gravity affects time as well, very soon.) Both space and time are relative: the measurement of each is affected by relative motion between observers. This is only bizarre because it is not part of our everyday experience, which is due to the fact that the speed of light is so huge compared to the "ordinary" speeds we experience. Still, there is no disputing the evidence that measurements of time are relative to the observer.

The second experiment involves taking one cesium clock to the top of a mountain for a few days, while leaving the other (identical, synchronized) cesium clock down below at the bottom of the mountain. We then bring the clocks together and observe how much time elapsed on each, *and the times are not the same*! Because of the precision of these clocks, we can measure even slight variations in time that we humans could not come close to perceiving under ordinary conditions. It turns out that the clock at the higher altitude ticks out time faster by exactly the amount predicted with Einstein's General Theory of Relativity. (As an aside, the Special theory involves objects moving at constant velocity, while the General Theory deals with acceleration or changing velocities). The conclusion is then unavoidable: gravity affects time in such a way that stronger gravitational fields cause time to pass slower than it does in weaker gravitational fields. So, at the top of mountains where gravity is a bit weaker than at sea level, clocks run faster than down below! We will explore this in much more depth when we delve into the four (at the moment) fundamental forces that govern all interactions in the universe.

Both the smallest and the largest time intervals we can measure, albeit not precisely, trace their way back to the universe itself. The largest time that has a significant basis in experimental evidence is the age of the universe, approximately 13.7 billion years old. This is deduced by measurements of the cosmic microwave background radiation (CMBR) present everywhere in the universe, as well as by determining the radial velocities of many galaxies (recall most of these galaxies are receding away from us at an accelerating rate) and extrapolating backwards to the point in time when *everything* was in one incomprehensibly dense speck of matter, energy, and spacetime. One might think of this speck as the granddaddy black hole of all time, or call it a "kugelblitz" for a black hole made from energy and not matter. Actually, we do not know how it got there at time zero or what it should be called. We do not know if there was a time before this point in spacetime came into existence, or even if the question makes sense. But the evidence nature provides is beyond dispute: the whole show began with a rapid inflation of spacetime from this speck of indescribable energy, and over billions of years evolved into galaxies, stars, planets, dinosaurs, and people. We know how to use the Doppler Effect and the relative luminosity of distant stars and galaxies to estimate distances and velocities, but how does this CMBR work to explain the age of the universe? Thus, we begin another adventure into the realm of cosmology.

In the 1960s, two astronomers working for Bell Labs in New Jersey, Penzias and Wilson, started working with a large, 20-foot radio antenna in the shape of a horn that was originally built to intercept radio signals from a satellite system called Echo, a name which turned out to be positively prescient. As Echo had been supplanted by a new and better satellite arrangement, Penzias and Wilson were given full license to explore the possibilities inherent with the antenna. If one looks through an optical telescope at the space between stars and galaxies, there is nothing but darkness in the range of wavelengths visible to the human eye. Such is not the case, however, if one uses a telescope tuned to the wavelengths of light associated with radio and microwaves. In that region of the electromagnetic spectrum, data in these supposedly empty spaces is readily available. It was this data that the two astronomers set out to study in what was to become one of the most momentous accidental discoveries in the history of physics. It is a classic example of making sure one pays attention to *all* the data nature is providing, and not just writing it off as "experimental error".

In analyzing the radio and microwave signals, they discovered an incessant hiss or noise or static coming from every direction in the universe, and in order to do the analysis they intended to do, they had to get rid of this background interference. And did they ever try! They developed systems to eliminate radio and microwave signals from urban sources, from the Milky Way itself, radiation produced by some stars, military signals, and finally pigeons. The horn antenna was so large that pigeons had taken to nesting in it, and so Penzias and Wilson cleaned out their droppings (and the pigeons). Nothing worked to eliminate this static. At about the same time, in what can best be described as extreme serendipity, a physicist named Robert Dicke from Princeton University was working nearby and had predicted the existence of a tiny remnant of radiation left over from the Big Bang. Penzias didn't know this of course, but the duo was so intrigued by this static that they called Dicke to tell him about it. It instantly became clear that what Penzias and Wilson had stumbled upon, quite by accident but also through painstakingly outstanding experimental work, was the electromagnetic "echo" of the beginning of the universe.

Roughly 380,000 years after the initial Big Bang inflationary period, the universe had cooled down enough for atoms of hydrogen and helium to form and remain intact, and since atoms are mostly empty space (yes, you are sitting on empty space), all the light that had been trapped in this primordial soup of particles and energy was free to escape and wander through the ever-expanding universe. Because of this continued expansion, there was yet another massive stretching out of space-time, resulting in longer and longer wavelengths as time progressed, until roughly 13.7 billion years later two astronomers detected it as microwaves, in an antenna occupied by pigeons. This was convincing evidence for the inflationary Big Bang model of the origins of the universe, and it also provided a map for cosmologists and astronomers to study regarding its evolution. Penzias and Wilson earned the Nobel Prize for this discovery in 1978.

We continue to analyze this CMBR today, and much progress has been made in our understanding of the structure of galaxies, their formation, and the expansion of the universe in general. All of this came from an annoying bit of static that perhaps most people would have finally written off as a nuisance to be neglected. If you owned a television in the 1960s, one of those with the "rabbit ears" sticking out from the set to receive signals from broadcasting stations, it is possible that this antenna arrangement would result in having the same static that was discovered about the same time. Imagine: convincing evidence for the Big Bang was right there on television, all the time!

The largest time interval we have measured, then, is just shy of 14 billion years. To put this in perspective, suppose at time zero the universe begins, and 2000 years later we arrive at the present moment. On this scale, the Earth establishes itself as a planet around the year 1300. Humans enter the picture around autumn of the year 1999, composed of atoms from the dust of stars that exploded eons ago. The smallest time interval we have to date is called "Planck time", in honor of Maxwell Planck, often considered the father of quantum physics. We will meet him shortly. For now, Planck time is calculated to be roughly 10^{-44} seconds, and can be determined by combining fundamental constants that permeate physics. (For the curious minds, the constants are G from the universal law of gravity developed by Newton, h from Planck's study of blackbody radiation and later Einstein's work on the photoelectric effect, and of course the speed of light c.) At this remarkably tiny slice of time, the laws of physics can (currently) go no further. Planck time is the amount of time it takes light to travel a Planck length in a vacuum. If we take the exact beginning of the universe as time zero, we only have evidence for what transpired from about 10^{-43} seconds and after. We do not now possess the laws of physics to go back in time any further than that.

So much of what we have explored regarding space and time seems to find its way back to the speed of light, and the very nature of light itself. It is fitting, then, to temporarily close this adventure with a foray into the history of our understanding of the characteristics of light. Once again, we will see that asking a simple question, such as: "what is the speed of light?" leads to numerous tributaries feeding into the main "river" that started the whole process. This last leg of the spacetime journey begins with the revolution in thought that brought us from a geocentric view of the universe to the current heliocentric system. This particular revolution took thousands of years to finally reach resolution; our understanding of what light is depends on it.

Does the Earth move? This fundamental question intrigued humans for centuries. Around 250 B.C. Aristarchus suggested that, in fact, the Earth does move around the Sun, and this model is coined the heliocentric system. But we have little evidence that his thinking gained any traction at all for a very long time. After all, look out your window: do you see the Earth moving? How could something as big as the Earth be moved, given that shifting the location of a large boulder takes a huge amount of force? There are objects in the sky that *do* appear to move. For example, the Sun rises in the East and moves across the sky to set in the West, and the Moon and planets move across the sky as well. It is reasonable to build a model of the universe where the Earth is at the center and everything revolves around it – the geocentric system. Moreover, it makes the Earth a central point in the universe, occupying a position of great importance. Hipparchus argued against the heliocentric model after calculating the orbits of the visible planets and discovering that they were not perfectly circular as it was thought they must be, and in the process figured out the distance to the Moon from the Earth using the basic geometry of triangulation, the method of parallax, and lunar eclipses.

Hipparchus was an extraordinary astronomer, perhaps the greatest ancient astronomer we have on record. His methods and calculations eventually wound up in the hands of Ptolemy around 200 A.D., and it is at this time that the geocentric model gained an extremely strong foothold not only in the realm of science, but also in the realm of established religious doctrine. The important thing to remember is that the Ptolemaic model was successful in many ways: it worked! Geocentric constructs can predict the seasons, the motion of planets, the Sun and Moon, and is (tacitly) used in all our terrestrial navigation. When we fly to Miami, for example, we assume the Earth is a more or less stationary reference frame. In addition, this system agreed well with Aristotle's teaching that the natural state of objects is to be at rest, and since it was thought that objects in the sky obeyed a separate set of rules governing their motion from the laws operating on Earth, there was no contradiction in people's minds.

In retrospect, it is easy to see why the geocentric world view held sway for many centuries. There was a major issue with it, however, and although explained away by Ptolemy by using complex circles upon circles called epicycles, the resulting system was quite convoluted. The problem was that some of the planets wandering through our sky at night appear to move *backwards* in their orbits around the Earth for a period of weeks at times. It was as though they stopped dead in their tracks, and then started going in the reverse direction for a while, and then decided to move forward again. Ptolemy figured out a mathematical way around this conundrum with his epicycles, but even the casual observer could see it was a bit contrived. What could make a planet move backwards? This problem is called retrograde motion, and it remained a complication of the geocentric model that was not easily explained away.

Centuries passed; the Dark Ages ensued along with Medievalism. In the 15th century the European Renaissance was ushered in, and with it a tremendous infusion of new knowledge and progress in art, exploration, and science. That set the stage for a Polish astronomer named Nicolaus Copernicus, who had the good fortune to have connections to the Catholic Church, as his uncle was a bishop. It is important to recall that at this time, it was perilous, indeed sometimes fatal, to go against the teachings of established religion, and the doctrine at this time rested squarely on the view that the Earth was at the center of the universe, with all "perfect spheres" orbiting around it. Copernicus had to proceed cautiously, regardless of his inside connections!

From around 1500 A.D. until he died at the age of 67, Copernicus developed his heliocentric model for the solar system, wrote a book titled *De Revolutionibus* containing these ideas, and received a copy of his book merely days before he died. In a sense, he was spared the years of turmoil that followed, because with the prevalence of Gutenberg's printing press (invented in 1440), the book found a much wider audience than would have been possible in ages past. Now there existed two competing models of the solar system, and most significantly the place of Earth and humans in each was vastly different in every conceivable way. Adherence to the Copernican model required people to cede the central (geocentric) position of Earth and its humanity in favor of a vast universe with Earth just being one of several planets orbiting the Sun. Clearly this was not a revolution in science, thought, and philosophy that was going to happen overnight. The debate between the two models raged on for one hundred years, with many other astronomers entering the fray.

Chief among the astronomers attempting to settle the debate was a man named Tycho Brahe, who lived from 1546 – 1601. Tycho's uncle was a vice admiral in the Danish Navy, and as he had no children, he asked Tycho's father if he might raise the boy or his twin brother as his own. The father agreed, and Tycho's twin was sent off to the uncle. Unfortunately, the tale took a cruel twist of fate when the twin died unexpectedly, and the uncle, feeling cheated of the deal, decided to kidnap Tycho in his stead. His uncle was quite wealthy, and together they lived in a castle while the uncle paid for his education. Though Tycho was schooled in diplomacy and the law, it appears a partial solar eclipse won him over to astronomy instead, much to his family's chagrin. He became obsessed with observations of the stars, planets, Sun, and Moon and began the necessary process in the settling of all debates and revolutions in science: experimentation. Without a telescope, since that had not been invented in his time, Tycho set about trying to prove that the geocentric model was the preferred system to depict the kinematics (the study of *how* things move) governing our solar system. He crafted beautiful huge sextants out of oak and brass to measure the angular position of the planets and stars, recording data so accurately that some of it can still be used today. His uncertainties in these measurements were remarkably small and were far superior to anything else at the time, and it is worth repeating *all of his data was gathered without the benefit of a telescope.* Tycho built underground rooms where he fashioned his observatories to protect his equipment from the weather, principally wind and rain. Eventually the king of Denmark gave him an island upon which to do his research, but things went awry quickly as Tycho became known as a wild eccentric, given to hosting raucous parties and designing his own dungeons when certain apprentices did not do his bidding. Tycho was forced to flee the island to escape the king's fury and, with his court jester, headed for Prague. He brought all of his remarkable data on the planets and stars, and there was a prodigious amount of it, with him. There are varying accounts of it, but at some point Tycho lost part of his nose in a sword fight, and for the remainder of his life wore gold and brass replacements for the missing part. It is not clear how Tycho died – it was well known he ate and drank way too much at banquets – but the thinking now is that it might have been a ruptured bladder, kidney stones, gall stones, or a failed liver. Whatever the case, the salient point is that all of his data fell into the hands of one of his apprentices, Johannes Kepler.

Kepler was an Austrian mathematician and a mystic by nature, having a profound belief in the supposed magic power of numbers and ratios. He lived from 1571-1630, and his role in the ongoing battle between the geocentric and heliocentric models was a vital one: organize the data that Tycho had so meticulously gathered. Most scientific revolutions follow a similar pattern: competing models emerge to explain patterns in nature, there follows a period of data gathering to ascertain which model is superior, people then organize the data to discern those patterns, and then one or more scientists puts it all together and (hopefully) settles the debate. Kepler spent over twenty years organizing Tycho's data, specifically his data on the motion of the planets. From this exhaustive study, Kepler deduced three planetary laws of motion. First, all the planets orbit the Sun in elliptical paths, with the Sun at one of the foci of the ellipse. Second, a line drawn from any planet to the Sun will sweep out equal amounts of space in equal times. A little thought on his second law indicates that planets must move faster when they are closer to the Sun in order for equal swaths of space to be swept out in equal times. Third, and perhaps the most fascinating, Kepler discovered that if one takes the mean (average) radius of orbit for any planet and cubes that number, then divides that cube by the square of the time it takes the planet to orbit the Sun once (called its period), one arrives at exactly the same answer regardless of which planet's data is used! It shorthand, the third planetary law can be written as $K = R^3/T^2$. Perhaps one can see why we use formulas in physics: it is so much easier than writing it all out! Recall that every formula tells a story, and this law is a most remarkable one, sure to excite Kepler because of his fascination with numbers and ratios. I doubt you missed it, but just in case, Kepler, somewhat ironically since Tycho was a firm believer in the geocentric model, had coined three laws which were clearly consistent with the heliocentric model of Copernicus.

While all of this is transpiring, the rest of the world is not sitting back and watching idly. Shakespeare is writing his great plays, Rembrandt is creating his marvelous paintings, there is a Protestant Revolt and the Church of England is established, the Renaissance fizzles out and the "new world", long inhabited by indigenous peoples, was accidentally bumped into by Columbus in his attempt to circumnavigate the globe. The British Empire gains strength, and amidst it all the telescope is invented during the lifetime of a genius born in Pisa, Italy, named Galileo Galilei. In the pantheon of physics, Galileo certainly is regarded as one of the greatest scientists who ever lived. The list of his achievements is astonishing, as was the story of his life, in large part due to those accomplishments.

I think we often teach great revolutions in science, philosophy, and most any field by citing the end result as though it was inevitable and obvious. This is a great mistake in our pedagogy, because it omits all the intricacies, experimentation, and human drama that creates this tapestry of nature we call science. By that omission, we risk people tearing at that tapestry with no evidence, failing to see that when one thread is pulled from this vast network of progress in thought, the entire tapestry falls apart, and one is left typing on a computer that should not exist in this mystic world of make-believe that people often conjure up, especially in difficult times. There is a wonderful book, written by James Burke, called *Connections*, in which it should become clear to any careful reader that all of human progress is linked in myriad and often unknown ways. One cannot choose to believe something to be false or fraudulent, usually because this will benefit the individual in some way, and then continue to utilize the very principles and objects that follow directly from the scientific truth they just chose to ignore! That is the essence of dogma and hypocrisy. It is important to recognize again that *belief is not necessarily truth,* and too often false beliefs and dogma lead us down paths of destruction.

At the most immediate level, we *believe* the Earth goes around the Sun, but most people do not know *why* we believe that – in a sense we have been brainwashed. It is also important to realize that the geocentric model is not wrong: it predicted a lot of phenomena in nature successfully, and even in the modern world there is much we do with the tacit assumption that the Earth is stationary. Galileo lived from 1564-1642, and the list of his contributions to the progress of science, and human history in general, follows. Though he was not the first to suggest that experience and experimentation was the best road to truth, he is widely considered to be the person responsible for implementing the scientific method of testing hypotheses and working toward repeatability of results under the same conditions. There are millions of creative variations on this "method"; there is no monolithic procedure to follow and there is no sense of final arrival at truth in any case. But every one of these variations, if it is to be tagged with the label of science, must rely on transparent, repeated experimentation at its core. Galileo *lived* this method by seeking the truths buried in nature's patterns through experiment, rather than belief and blind obedience. He rolled objects down inclined planes and eventually came to the conclusion, contrary to Aristotle's teachings, that the natural state of objects is *not* to be at rest but rather to *resist changing* the state of the motion they are currently in, so that an object moving without any friction or resistance would continue to move in the same manner forever until made to change that motion by an external agency. This is the famous law of inertia, which still has no fundamental explanation: we do not know why objects have this tendency to maintain their velocity, we just know they do!

Galileo studied projectile motion, such as throwing a ball through the air, and astutely observed that horizontal motion is entirely independent of vertical motion. He challenged Aristotle's notion that heavier objects fall faster than lighter objects *by testing this idea*. It is reported that even after dropping objects of vastly different weights and proving they landed simultaneously, some "learned scholars" still clung to Aristotle's hypothesis (which had never been fully tested until Galileo came along). Therefore, Galileo was the first to show that the acceleration of gravity, the way objects speed up when released near the Earth, was exactly the same value for all objects, no matter the weight of the object, and assuming any frictional forces such as air resistance were not a factor. He developed the physics of the swinging pendulum, and improved on the newly invented telescope with his own modifications that were to later prove so useful. It's this last piece that we will focus on now, as it leads us back full circle to the debate regarding competing models of the solar system and the place of Earth in the cosmos. It also, finally, will bring us to the point where we can start answering questions about the nature and speed of light, because *now* we will have the proper historical and conceptual context in which to do so.

With the telescope, Galileo was well equipped to turn his attention to the planets and stars. He observed moons orbiting Jupiter, Sunspots, and craters on the Moon. All three were anathema to the geocentric model: how could some objects move around Jupiter if everything goes around the Earth, and if these "perfect spheres" being discussed have sunspots and vast craters, what else might be wrong in that scenario? Though some people persist in believing things even when their own eyes tell them it's otherwise, most humans (hopefully) will lay their beliefs down when irrefutable truth through observation and data is placed in front of them. Galileo thus became a major, forceful proponent of the heliocentric model, and in the process the retrograde problem concerning the motion of the planets in our sky was laid to rest with a very simple explanation. Since the Earth is moving around the Sun faster than Mars or Jupiter (or any planet farther from the Sun), there will be times from our vantage point where we "pass" those outer planets and they begin to appear to move backward in our reference frame. This is so much simpler and more logical than inventing fancy ways (epicycles) to explain this apparent motion, and therefore the Copernican model, from the outset, is a superior model for its simplicity, its ability to predict all the things the geocentric model does, and for its success in explaining a few more observations that are not predicted and could not occur in an Earth-centered system. I will focus on one of these items, and then we can rest our case as to *why* we are certain that we live in a vast universe as a speck of blue orbiting around a very ordinary, main sequence star that gives us life as we know it.

This item is termed "stellar parallax", and we will get to it shortly. But before we leave Galileo, there is one more scientific item to put on the table which was used marvelously well by Einstein centuries later: Galileo posited that all reference frames having constant velocity, which means no change in speed or direction and thus includes staying at rest, are in every way equivalent! As an example, if we are moving in a vehicle at constant velocity, there is no experiment we can do in that reference frame that will tell us we are moving. If we flip a coin vertically, it comes back into our hand just as though we are at rest. If we hang a pendulum, it hangs down vertical, just like it was at rest or at any other constant velocity other than zero. If we drop a jackknife from the top of a mast of a sailing vessel moving at constant velocity, the knife will strike the bottom of the mast just as though we are at rest. *There is no absolute frame of reference: all constant velocity reference frames are equivalent.* It may not seem profound, but it is a deep insight into nature that has huge implications when we discuss the speed of light and the very essence of space and time. In this sense, every object in the universe is moving relative to some reference frame somewhere. There is no spot in the universe that we can call "special" or absolutely still. I cannot leave this brief sojourn with Galileo without paying homage to his remarkable courage in the face of ridicule, persecution, and outright imprisonment. I urge the reader to investigate the full scope of his life – there are many books which do so wonderfully. He is unquestionably a giant of physics, but I think he also deserves a place that occupies a much wider kingdom: those rare individuals who transformed the human species at considerable cost to themselves.

The analysis of stellar parallax will bring this scientific revolution, culminating in our adoption of the heliocentric model, to a close. The simplest way, perhaps, to think of stellar parallax is to picture the Big Dipper. We might expect that the *shape* of this constellation would alter somewhat as the Earth moves around the Sun, since our vantage point is constantly changing. The stars comprising the Big Dipper should apparently shift their position against the background of stars that are farther away if the reference frame they are viewed from is moving. Ironically, this phenomenon was a feather in the cap of the geocentric model, since clearly it does not appear that the Big Dipper changes shape at all! We are not describing how constellations move about in the sky here – that has to do with the spin of the Earth (on any given night) and the orbit of the Earth around the Sun (which is why constellations vary through the seasons). For the most part, the movement of stars in the nighttime sky can be explained by either the geocentric or heliocentric model, but the relative position of the stars is another matter. Therefore, the *apparent* lack of any stellar parallax for many centuries was difficult to explain using the heliocentric model, unless of course the distance to the stars is unimaginably greater than was previously thought.

And that is, in fact, the case. Recall that on our trip through the universe at the speed of light, it was a matter of some hours to escape our solar system, but it took *years* to reach even the nearest star to our Sun. Think of viewing a triangle of cones from a short distance. Now as you move around the cones, notice how the shape of the triangle appears to change – parallax! Now imagine you are gifted with extraordinary vision, move hundreds of miles away from the cones, and move around them in orbit again. Can you visualize how the cones (stars) are going to look about the same? Nonetheless, even at the mammoth distances of the nearest stars, there should be a slight stellar parallax due to the Earth's motion around the Sun, and with the advent of newer and more precise telescopic observations, it was early in the 19th century when this parallax was accurately measured and the matter was resolved. There is no disputing that the Earth moves relative to the distant stars in its orbit around the Sun. It is important to note that no-one, not even Galileo, has proposed the *mechanism* that causes the Earth to move around the Sun in a way that is verifiable and predictive. That puzzle was left to Isaac Newton, the man who stood on the shoulders of the giants we have discussed previously, and who was born the same year Galileo died (1642). We will get to know Isaac in much detail, but for now, the stage is set for the final phase of space and time exploration: the speed and nature of light itself.

What is light? How does it interact with matter? How can we measure its speed if it is infinite? These are among a myriad of questions about the nature of light that have intrigued humans for centuries. Poets, philosophers, artists, and many others have proposed thousands of ideas regarding light and its behavior for millennia, but the first real scientific breakthroughs in terms of how light behaves came to fruition with the work of Isaac Newton. He explored how visible white light can be broken down into its various colors, and postulated light to be a stream of particles, the "corpuscular" model as it was deemed by some. One of his contemporaries, Christiaan Huygens, proposed a wave model of light based on some other observations. This wave-particle dichotomy, unbeknownst to physicists at the time of Newton, was destined to chart the course for one of the most successful theories in physics, quantum mechanics, and drove the discussion for the next three hundred years. Like all scientific revolutions, when two competing models vie for acceptance, once the stage for "battle" has been set, the road to resolution must at some point include extensive experimentation. Witness the many centuries it took to finally resolve the debate between the geocentric model and the heliocentric model! Around the year 1700 AD then, the study of light began with the competing models of particles versus waves, and it was left for those who followed Newton to attempt to resolve the dispute.

The first major step in the resolution as to the nature of light occurred around 1800 AD, when Thomas Young performed his famous double-slit experiment. In this experimental setup, monochromatic light (one color = single wavelength) was shown through two very narrow slits that were separated by a small distance. These slits were often only millimeters wide, and the separation between them only a centimeter or so. Young observed the pattern this single wavelength (note how we are already trapping ourselves into the language of waves!) light made on a distant screen, and it appeared as a series of "fringes". If one drew a line from the midpoint between the slits to the screen, that central location on the screen had a bright fringe – it looked like a small, rectangular block of bright light. Right next to it, symmetrically on either side, there were regions of total darkness or no light whatsoever. As one moved outward away from the central point, this pattern of light-dark-light-dark fringes kept repeating, though the farther from the central point one got, the dimmer the bright fringes became. Young argued persuasively that these fringes were conclusive evidence of interference, with the bright fringes showing constructive interference (places where the two light waves added up to a bigger wave), and the dark fringes destructive interference (places where the two light waves canceled each other out). Thus, he argued, light must behave as a wave and not a particle, because only waves interfere; particles do not. Apparently, he reasoned, the great Isaac Newton was wrong for once.

The second major step, also occurring in the 19th century after Young's famous experiment, was the outcome of the historic Poisson-Fresnel debate. Fresnel was an advocate for the wave model of light. Poisson was highly skeptical, and proposed an experiment that he thought would predict a preposterous result based on the wave model of light. (I have done this experiment with a laser, and the results are astounding!). Poisson set up the experiment so that monochromatic light, accomplished back then by filters, was to be shone on a small, circular object that is opaque to light. He then predicted, again in a sarcastic manner, that if the wave model of light was correct, the shadow of this circular object, seen on a distant screen, would have a single bright spot in its center, since the diffraction (how light spreads out and around a barrier) from the outer edges would have the light travel equal distances to that center point, thus arriving in phase as constructive interference, thus producing a bright spot that has come to be known as "Poisson's Dot". His argument was precise and correct; he thought the result to be patently absurd. The game was on, and in an historic, public demonstration, the experiment was done. Lo and behold: there was (and is) indeed a bright spot at the center of the shadow! The wave model of light now gains immense traction.

The third critical piece in the progression of ideas about light was encapsulated by the work of James Clerk Maxwell. You will read in a future section how he codified all of electromagnetic theory into four laws known as Maxwell's Equations. Here there is a description of how charge produces an electric field, how magnetic fields interact, how electricity can produce magnetism and how changing magnetic fields can produce electricity. But the key element that Maxwell added was that accelerating electric charge can produce an electromagnetic wave, and he calculated that the speed of this wave was, in fact, the speed of light! Therefore, the wave model of light gains still deeper consensus in the physics world, and experiments by Tesla, Hertz, and Marconi confirmed that generating radio waves is possible. So, it appears that light does travel as a EM wave.

The fourth major step occurred toward the end of the 19th century, and was a logical outcome of the widespread acceptance that light behaves according to the wave model. Since all waves that we know of require a medium to be transported (one cannot think of water waves without the water to make them go, or sound waves without air to allow them to propagate), it was reasonable to assume there must be a medium in the universe, *everywhere* in the universe, that allowed light waves to travel from one point to another. This medium was dubbed "the ether", and the search began to detect this ether experimentally. Michelson and Morley, two scientists who worked together on this project, embarked on this exploration by using an ingenious device called an interferometer. Interferometers work by sending light waves along two perpendicular, usually equidistant paths and, by using mirrors, merge these light waves to make an interference pattern upon their return.

Their hypothesis was that if the ether is everywhere, then in order for light waves to be transported between the Earth and the Sun, there must be that same ether between the two bodies since sunlight clearly reaches us. The Earth moves around the Sun (you see how and why the heliocentric model is vital here – everything connects!) about 66,000 miles per hour relative to the Sun. Michelson and Morley thus reasoned that the Earth should encounter an "ether wind" as it travels through it. They set up their experiment with the interferometer so that one light beam traveled into the ether wind, while another light beam either traveled perpendicular to it or with the wind. The theory was that the light waves traveling with the ether wind would move faster than the other light waves moving against or perpendicular to the wind, and that this would be detected by a shift in the interference pattern. Even though 66,000 miles per hour is very slow compared to the speed of light, it is fast enough to easily be detected by a sensitive interferometer. The experiment was repeated many times with many orientations, and always the results were the same. It is the most famous null result in the history of physics: there is no ether and none has ever been detected. It appears the speed of light is constant to all observers regardless of relative motion or reference frame. It is also evident that if light is a wave, it is a very strange one indeed, because it requires no medium to be transported. And so, the wave model of light by the year 1900 was still intact, but the results of this experiment remained a deep puzzle.

In the year 1900, a conservative German physicist named Max Planck was doing his graduate work and decided to study blackbody radiation. This phenomenon is essentially a study of how a lump of coal radiates away light energy. The experimental results agreed with Newton and Maxwell to a point, but the data involving higher frequencies in the violet and ultraviolet regions was wildly different from the predictions of classical physics (Newton and Maxwell). It was clear something was terribly wrong about this classical theory, and Planck set about trying to solve what was deemed the "ultraviolet catastrophe". In the final analysis, he was only able to do this if he assumed that the light being emitted by the blackbody only came in discrete bundles of energy called quanta, with each bundle having an energy of hf. The constant "h" was dubbed Planck's constant, and had a value of 6.626×10^{-34} Joule•secs, and "f" is the frequency of the light being emitted. These bundles of energy had energies of 1hf, 2hf, 3hf, and so forth but never something in-between. This avoided the ultraviolet problem and matched the data, but Planck was reluctant to assign any real significance to the idea of quantized energy. Little did he realize that he had ushered in the birth of quantum mechanics.

The year 1905 is often called the "magic year in physics", because an obscure patent clerk submitted several papers for publication that year, the sum total of which laid down the mainlines of physics for the next 120 years to the present day. One paper dealt with the motion of suspended particles in stationary liquids, based on a molecular viewpoint, which lent convincing evidence as to the existence and behavior of atoms. This harkens back to Lucretius observing that dust particles suspended in air undergo a random dance when sunlight streams through, and also provided an explanation for the so-called Brownian Motion observed by the botanist Robert Brown as he looked at pollen grains suspended in liquid under a microscope in the 19th century. Two more papers presented the Special Theory of Relativity (STR – but the papers did not go by that title) in which space, time, and mass were seen as *not* absolute, but rather dependent on the reference frame of the observer, and mass or inertia was related to energy content. In these remarkable submissions, the speed of light was seen as the absolute – the same to all observers regardless of reference frame.

The work drew from the ideas set forth by Maxwell and (perhaps) from the results of the Michelson-Morley Experiment, in which no ether was found and the speed of light was the same in every instance. One of these papers saw the birth of perhaps the most famous equation in all of physics: $E = mc^2$. Most observers would agree that these papers were met with significant skepticism and, in some cases, outright scorn. Who was this patent clerk talking about time slowing down, space contracting, and mass increasing as speeds approach the speed of light? The technology to test this theory was not readily available at the time, but in the last several decades relativity theory has been proven to be correct in millions of experiments and applications. Recall the cesium clocks, GPS, and the presence of an abundance of muons at sea level we have explored. These are just a few examples verifying the two postulates of STR: all constant velocity reference frames are equivalent (Galileo) and the speed of light is the same to all observers, regardless of reference frame.

The fourth paper is the one we will focus on, and it dealt with a phenomenon called the "photoelectric effect". It was this paper that won the patent clerk a Nobel Prize 16 years later in 1921. Of course, the unknown patent clerk was Albert Einstein. In the photoelectric effect experiment, monochromatic light (one color, one wavelength, one frequency) is shown on a metal surface in a vacuum tube so that no air is present to interfere with the results. The idea was that since electrons in metals are loosely bound to the nuclei of the metal (which is why metals are good conductors), perhaps enough light energy could liberate the electrons from the metal and eject them off the surface to be collected by a positive anode. A circuit could then be set up that would detect these ejected electrons, and even measure their kinetic energy (KE) by using a "stopping voltage" in the circuit. The concept was very straightforward: if light is a wave, then the intensity or brightness (think amplitude) of the light should be directly correlated with the KE of the ejected electrons. Simply put, brighter light should knock electrons off the metal with increasing KE. An analogy is useful here: think of an ocean wave slamming onto the shore. The height of the wave (brightness/amplitude/intensity) should knock small pebbles (electrons) laying on the beach (the metal) off the beach with large KE values. We would expect that a 20-foot wave would smash the electrons/pebbles off the beach with huge KE values. But that didn't happen, ever. It turned out when red light was used for most metals, no matter how bright the light was made, the electrons just sat there on the metal and did nothing. According to the wave model, a 20-foot wave has no effect on the pebble sitting on the beach. Clearly there is a big problem with this model!

The plot only thickens from that point, because it was then found that when even very dim blue light was shown on the metal, electrons came flying off the metal with large KE values! If the wave model is to be believed, the height of the wave has no effect, but the spacing of the waves (frequency) is what causes the electrons to be ejected. This is akin to positing that a 20-foot wave does nothing, but if the waves have the right spacing apart, then even a 2-inch wave would smash the electron/pebbles off the metal/beach. In other words, the wave model makes no sense and is an absolute failure here! Einstein showed that when the KE of the ejected electrons is plotted against the frequency of the light used, a linear relationship was seen, and the slope of this line was Planck's constant. Clearly what Planck had stumbled upon was far more than a small curiosity. Yet from a particle standpoint, the results make perfect sense if we consider light to be particles called photons, each with energies of 1hf, 2hf, 3hf, and so forth. The results are entirely consistent with the energy of the light being related to frequency, not intensity!

Confusion ensues. Young's double slit, Poisson's Dot, EM waves…all seem to indicate light is a wave. Yet here is conclusive evidence that light is a particle. How can it be both? Experiments are proposed for the double slit apparatus that Young used, wherein a detector is placed behind the two slits. The idea is to trick the light: if it is a particle, we can see which slit the photon came through and know it is a particle. Indeed, we can do this, and when we do, we can detect which slit the photon comes through, but just as we are able to do this, the interference pattern on the screen disappears! And just when we "weaken" the detection system to the point where we can't tell which slit the light comes through, the interference pattern re-emerges. It appears that when we look, light is a particle, but when we don't look, it acts like a wave. Welcome to the weird world of quantum physics!

It is difficult to convey the complete revolution going on in physics during this time, but here is a sampling. A significant portion of the discussion which follows will be explored in depth in future sections, if it hasn't been already, but for the sake of clarity regarding the study of light, we will visit them briefly at this juncture. With the advance of high voltage technology and

EM theory in the late 19th century, along with glass-blowing and vacuum tubes, it became possible to generate cathode rays by "boiling electrons" off the cathode at high voltage and collect these at the anode inside a vacuum tube. This is the primitive form of television! By using magnetic fields, we found these cathode rays were negatively charged, hence electrons were born. Roentgens and others put targets in the way of these cathode rays, and found new penetrating rays coming out the other side of the target: X-rays! Becquerel discovered penetrating radiation came from uranium deposits, and Madame Curie further investigated this, introducing alpha particles and the element radium (the ideas of radioactivity and nuclear decay were soon to follow). Rutherford and others used these alpha particles and fired them at a thin gold foil, with the results establishing a positively charged, dense nucleus and ushering in the Bohr planetary model of the atom with electrons orbiting the positive nucleus. This model had some success explaining the spectrum of hydrogen gas placed under high voltage, but could not explain why the emission lines in the spectrum varied in intensity, exhibited "fine structure" splitting of lines, and the model failed when applied to more complex atoms. Millikan found the value of the elementary charge on an electron using oil drops, and de Broglie hypothesized that *all* matter has wave-particle duality, not just light, and that the wavelength of matter was equal to h/mv (Planck's constant divided by mass x velocity or momentum). Einstein extended his relativity to the General Theory involving acceleration and a completely new way to think of gravity as the distortions of space-time produced by mass. This remarkable theory is still being verified today as gravity waves, generated many millions of years ago by colliding black holes or neutron stars, have finally been detected as tiny ripples here on Earth by LIGO facilities (to be discussed shortly). Compton showed that light particles called photons scatter off electrons much like the collisions between billiard balls. Hubble showed that the Doppler Red Shift in the spectrum of starlight indicated many stars were made from hydrogen, and that the universe must be expanding! Meanwhile, Schrodinger and Heisenberg refine the ideas of quantum physics, and cast electron probability clouds and uncertainty as principles built into nature, serving as the foundation of quantum mechanics. The neutron was discovered by Chadwick in the 1930s, and fission and fusion were soon to follow, with immense implications for the entire human species. This is just a sampling of the incredible synergy occurring during this time in physics.

So, where does all this leave us with regard to the nature of light? A theory was developed by Richard Feynman and others in the late 1940s and after, whereby all the physics of how light interacts with matter was codified into the particle model of light called Quantum Electrodynamics, or QED for short. This has turned out to be the single most successful theory in all of physics, predicting experimental results to an unbelievable level of accuracy and precision. I refer the reader to a small book titled *QED, the strange theory of light and matter*. It is an incredible foray into the nature of light.

The study of light is typically broken down into geometrical (ray) optics and physical (wave) optics. In geometrical optics, the way light reflects off boundaries and surfaces, as well as refracts (bends or changes direction) when entering into a different medium, is explored. This gives us the physics necessary to describe how mirrors and lenses work, with all the attendant applications such as telescopes, microscopes, corrective lenses and much more. In physical or wave optics, light is studied from the wave model perspective, and various interference patterns produced by two (or more) thin slits are analyzed, along with diffraction and some other related phenomena. The section of the book involving the four fundamental forces in nature will dig further into some of this material, but for now we will return our focus as to how light itself became deeply entrenched not only in how we think of space and time, but also how we *measure* both. This final piece brings us to the discussion as to how we figured out the value for the speed of light, which you may remember was part and parcel of how we defined the amount of space called one meter, which in turn involved the period of time called one second.

Galileo made a gallant attempt at measuring the speed of light. As the story goes, he and a friend were poised on top of separate hills a measured distance apart, and using his pulse as a timer, Galileo would clock how long it took light to go from him to his friend. They attempted this by lifting the shutters of their lanterns, first when the signal was sent and then again when the light signal was received. It didn't work. The only conclusion that could be drawn was either the speed of light is infinite or it is extremely fast! Recall if we could send a light beam around the entire planet at the equator, it would make several complete orbits in one second, so there was no possible way Galileo's method would be measureable: human reaction time alone would have made it impossible. Yet it wasn't long after Galileo died that the first, reasonably accurate measurement of light speed was obtained.

If you decide to look this astronomer up and do some background reading regarding his work, be prepared that his name is spelled in many different ways – something like how the same character in *War and Peace* goes by so many names that it is easy to get confused! His last name has three variations that I came across: Roemer, Rømer, and Romer with an umlaut over the "o". His first name also has three variations that I found: Ole, Olaf, and Olaus. I'll leave it to the reader as an exercise to come up with all the possible combinations from the three choices in each! For now, I am using Roemer, and the time is around 1676 when he began to study the eclipses of the moons of Jupiter. There are a few notable items embedded in the previous sentence: the telescope had to have been available, Galileo had made it well known that Jupiter had moons, and a more accurate method of measuring time (other than someone's pulse) was employed. It makes sense to use astronomical distances, since the speed of light is so enormous. I am going to simplify Roemer's method, without changing the essence of how he came up with a value for the speed of light. I refer you to an excellent book I came across, thanks to a brilliant and kind friend of mine from long ago, titled *Fundamentals of Optics* by Jenkins and White – this book is a rich resource far beyond a complete analysis of Roemer's method. So, thanks Walter for giving that book to me; I still read the notes you made amongst its pages and I still watch for that redtail hawk circling in the evening, wondering if that could be you.

Io is one of the inner moons of Jupiter, and observation indicates it takes 42 hours, 28 minutes, and 16 seconds to complete one revolution around Jupiter. How does one make this observation? There are several ways to do this, but Roemer chose the method whereby, looking through the telescope, the time between the consecutive emergences of Io from Jupiter's shadow was measured. (One could also measure the time of passage between passing *into* the shadow of Jupiter). This is a very straightforward observation that can be done with a reasonably powerful telescope, available during his time, and the period of Io's revolution is constant and quite predictable. Therefore, and this is where I will simplify the scheme a bit, suppose we do the following, noting well that this entire analysis relies on the heliocentric model. There are two times in our orbit around the Sun when the Sun, the Earth, and Jupiter line up in exactly a straight line. In the near point situation, a line drawn from the Sun passes through the Earth and then out to Jupiter: this is when the Earth is closest to Jupiter. In the far point situation, the line starts from the Earth, passes through the Sun, and then out to Jupiter: this is when the Earth is on the opposite side of the Sun from Jupiter. One might think that this just means six months from near point to far point, since it takes twelve months to go around the Sun, but we must remember that Jupiter is moving around the Sun also, albeit much more slowly, so the actual time is a bit more than six months before the two planets are in opposition. We know the exact locations for these points and also the exact times they occur, so we are now set up to do the experiment.

At the near point, we note the time that Io emerges from Jupiter's shadow, and from that observation we can predict when the next emergence of Io should take place *at any time* in our orbit around the Sun, because we know the relative positions of the planets at all times and we know the period of Io does not change. The easiest location to use, other than the near point, is the far point, and since we know how much time will elapse for the Earth to get to the far point from the near point, we can predict with only a small degree of uncertainty when Io should emerge from Jupiter's shadow a little over six months from the first (near point) observation. This is not exactly what Roemer did, but it is much the same idea, because if we do this experiment and wait expectedly at the far point for the precise time Io should re-emerge from Jupiter's shadow, we find that it is delayed by 1000 seconds or so from our prediction. This is not some sort of error in measurement – we would never expect to be off by that amount of time given the fact that Io's period of revolution is very consistent. Roemer was a smart astronomer, however, and he knew the delay in seeing Io emerge from Jupiter's shadow was because at the far point, the light carrying that information had to travel an extra distance equal to the diameter of Earth's orbit to reach our eyes compared to when we first observed the emergence at the near point. Had he known the correct distance of the Earth to the Sun, he would have nailed the value for the speed of light! The correct (mean) value for the Earth's radius of orbit is 93 million miles, so the diameter of our orbit around the Sun is 186 million miles. Dividing the extra distance by the delay time to get the speed, we have the speed of light equal to 186 million miles divided by 1000 seconds, which give 186,000 miles per second – very close to the accepted value today!

As the years passed, more and more precise measurements for the speed of light were made, using the aberration of starlight, rotating mirrors here on Earth, microwaves in a hollow resonator, and several other experimental methods. At the moment, it is laser light we use to obtain the value for the speed of light as 299,792,458 meters per second in a vacuum. You may remember this number: we used it to define the meter and by extension the second. The speed of light is universal, constant, and available to anyone who lives in our universe, and it is therefore the natural metric of spacetime as we know it. *All of kinematics*, the study of how fast, how far, how much time it takes things to move through space, boils down to meters and seconds. We now have a standard of measure for the meter and the second that is incredibly precise, reproducible, and indestructible. Along the way we have also discovered that space and time are inextricably linked, and in fact cannot be separated in terms of actual measurements. This brings our discussion of space and time to a temporary conclusion.

MASS

Feynman once wrote that if all of the scientific knowledge in the world were to be obliterated in some calamity and we had to choose just one fact to save in the process, he thinks it should be the idea that everything is made up of atoms. The atomic hypothesis is at the core of any discussion of mass, and the idea of indivisible particles making up matter goes far back in our history. Democritus, around 500 B.C., raised this proposition, and he wasn't the only one in all probability. Later on, in the first century B.C., the Roman poet Lucretius presented the first empirical evidence for atoms that we have on record by observing that sunlight coming through a window can allow one to see dust particles in suspension, and that these particles dance about in random directions. He suggested that the cause of this random motion for the dust must be invisible particles slamming into them, and in so doing was nearly 2000 years ahead of his time.

The word "atom" originated from the ancient Greek language and translates literally as "uncuttable" or "indivisible". As is the case with most ideas, the atomic hypothesis evolved along with its creators. It was thought that atoms were indestructible, indivisible, different in size and shape, continually in motion, and able to form various substances by combining in different ways. A fairly coherent theory of mass was developed which could partially explain the variety and change we see around us, despite the fact that this theory had little predictive power. Yet the atomic hypothesis seemed to take a fatal step when it was proposed that atoms exist in empty space and are able to cross these empty spaces to interact with one another. Without proof, without the formal structure of the scientific method and an advanced technology, the atomic hypothesis could be dismantled with rational, logical rhetoric by a gifted orator: Aristotle. He is famous for claiming "nature abhors a vacuum", and when he was finished beating this fragile idea of atoms into the ground, it remained imprisoned and seemingly lifeless for nearly 20 centuries. Perhaps we shouldn't be too hard on Aristotle, as one could make the argument that humanity was not ready for the immense power this idea can unleash, and he gave us much to think about regardless.

No idea, scientific or otherwise, is ever advanced in a vacuum. There are political, economic, cultural, and religious pressures to be taken into consideration. The atomic hypothesis had its genesis in a culture that was advanced in many ways, but we also need to recognize that there was a concurrent, deeply rooted belief in supernatural "gods" who supposedly took control of natural phenomena in a manner far beyond the means of mere mortals. Imagine being the first person to witness a violent thunderstorm – you might easily think the world was coming to an end or that some supernatural being in the sky was angry for some reason. Think of volcanoes, earthquakes, and hurricanes. Who would not be terrified of these events when they first were witnessed, with no explanation and no assurances? It is reasonable to assume that most people will not react well to any idea that claims to strip power away from these gods, for to accept such a notion risks incurring their wrath in retribution of man's arrogance.

We know from painful experience that for every person who attempts to open a door to the future, a thousand people are appointed to guard the past and keep that door closed. The Greek culture produced an amazing array of ideas that run the gamut from science and philosophy to organization and governance, but it remained bound to superstition and dogma simultaneously. In the midst of this remarkable civilization, Aristotle emerged as a formidable opponent in the art of rhetoric. His arguments were constructed in a logical manner, and his ability to cast opposing ideas in doubt was unrivaled.

Aristotle proposed his theory of matter around 350 B.C., drawing upon past ideas from Empedocles suggesting that there were four fundamental constituents for all mass: Earth, fire, air, and water. These basic building blocks could be combined in various ways to account for everything observed in the universe, but when it came to the "atomists" such as Democritus, there was a distinct parting of ways. Aristotle argued that if one continued to chop an apple in half, there would never be a point whereby an indivisible particle was reached – matter must be continuous. The atomists had a nearly impossible task in front of them: one cannot see atoms, scientific methodology was many centuries away, and the proposition of empty spaces between atoms was anti-intuitive to most observers. Aristotle added qualities he labeled hot, cold, moist, and dry to the four elements of Empedocles to outline a consistent theory of matter that supposedly could account for all types of substances. Herein lie the ancient seeds of alchemy and chemistry! Aristotle's arguments carried much weight, and absent any proof on either side of the debate about atoms, his skilled rhetoric appealed to the majority of people in his day, people who were fearful of the unknown (and the gods), and who yearned for the security of the spoken word that plays to their desires with tantalizing logic. This is a theme that never seems to wither away in human history, and it is one that too often, ironically, leaves chaos and destruction in its path.

The atomic hypothesis thus lay dormant, but not dead, for many hundreds of years. It has been roughly translated that Einstein once said "imagination is everything". We will never know if it was fortunate that our imagination failed us way back in ancient Greece, or where we would be now (if at all) had the ideas of science and atoms ruled the day. As the centuries rolled by, the ideas of Dalton, Priestly, and Mendeleev unfold, and a botanist named Robert Brown, who we have met briefly many pages ago, begins his study on pollen grains using the developing technology of the microscope. Underneath magnification, these pollen grains are observed to execute a "random walk", zig-zagging about in unpredictable ways, much like the dust particles Lucretius observed eons ago. The atomic hypothesis now slowly awakens from its deep sleep, and Boltzmann, with his work on statistical mechanics, and then Einstein through his analysis of Brownian Motion, bring it to the front burner of scientific inquiry. The idea that all things are made from atoms now gains a firm foothold, and no amount of logic or rhetoric can dissuade people from the gathering evidence that the atomic model works to explain literally millions of processes in nature. The challenge now becomes to identify the structure and substance of these atoms: what do they look like and what are they made from?

We enter the early part of the 20th century, at which time Madame Curie has discovered that certain rocks produce a penetrating radiation, dubbed alpha particles, that through electromagnetic experiments prove to be positively charged. We will investigate electromagnetism in detail when we explore the four fundamental forces, but for now we use the fact that alpha particles exist in nature and that they possess a property of matter called charge, which is the last of the four concepts undergirding nearly all of physics, and the one that we will delve into next. Ernest Rutherford, with the help of his able graduate assistants Geiger and Marsden, designed a clever experiment in an attempt to deduce the structure of the atom.

Imagine building a large wooden box with a cover on it so you cannot see inside. Somewhere inside this box a small triangle is built with nails and rubber bands. Students are allowed to roll marbles through this box, but cannot look inside. Their task is to deduce the size and shape of what is inside by observing what the marbles do when they pass through the box! I have done this with my students many times, and while they have a lot of fun doing it, they also see it is difficult to solve the puzzle. Rutherford used the same idea to discern the nature and structure of what is "inside" the atom. Instead of marbles, he used alpha particles, and he sent these positively charged particles through a very thin gold foil. Surrounding the gold foil was a detector, and each time an alpha particle hit the detector a tiny flash of light could be seen. During the actual running of this experiment, two competing models for atomic structure were vying for acceptance. One hypothesis was called the "plum pudding or blueberry muffin model", set forth by J.J. Thomson and others, in which negatively charged particles (the plums or blueberries) were interspersed among the positive matter (the pudding or muffin). The other model had the positively charged matter in the center with the negative particles circling around it, and this was (understandably) called the planetary model. At the time, the plum pudding model seemed to be the favorite among scientists. Once again, rhetoric and belief hold no sway in science: experiment provides the answer, and in this case, the results were a giant leap forward in our understanding of matter.

First, nearly all, upwards of 99% of the alpha particles, went straight through the gold foil undeflected. Reflecting back to the marbles and the wooden box, one might conclude that the box is completely devoid of any structure inside since the marbles pass through without any deflection. Further investigation of the results, however, reveals two patterns, one of which was truly astounding to Rutherford. A small percentage of the alpha particles were deflected at appreciable angles away from the original line of approach, and (remarkably) a tiny percentage were scattered nearly *backward* from the target foil. In Rutherford's words: "...as incredible as if you had fired a 15-inch shell at a piece of tissue paper and it came back and hit you!" The only way to explain these results, quantitatively and qualitatively, was to picture the atom as mainly empty space (the reason for most of the alpha particles passing through undeflected) with an extremely dense, positively charged nucleus at its core. This last conclusion follows from the fact that like charges repel, so when a positive alpha particle happens to "collide" with the nucleus of a gold atom, it gets repelled backwards out of the foil without ever actually touching the gold nucleus, *assuming the nucleus is charged positive also*. Since these events are exceedingly rare, the nucleus must be very small compared to the atom – about 100,000 times smaller when the numbers are crunched – hence with such a small volume the density of nuclear matter must be enormous, around 10^{15} grams per cubic centimeter. If a small cubic centimeter of sugar had this density, it would weigh somewhere around a billion *tons*! In fact, the radius of the nucleus is on the order of femtometers, or 10^{-15} meters, whereas the atom itself is on the order of Angstroms, or 10^{-10} meters. If we pictured the atom as a football field, nearly all of its mass would be located at the 50-yard line and would be about the size of an orange, with the negative particles swirling about the endzones and the rest completely empty space!

At this point the planetary model wins the debate, and the atom is visualized as having a dense, tiny, positively charged core called the nucleus, while being surrounded by negative particles in "orbit" since opposite charges attract. Of course, we now know positive charges in the nucleus are called protons, and the negative charges in orbit are the electrons. Since atoms are normally electrically neutral, the amount of charge on the proton must be balanced by the same amount on the electron. The exact amount of this charge was experimentally determined by Robert Millikan, about the same time as Rutherford did his experiment, and we will delve into this in our exploration of the fourth basic concept, electric charge, shortly.

The pieces of the puzzle now start to fit together, and a picture emerges as to the structure and nature of matter. With the advent of electrical power in the 19th century, high voltage could be applied to metal plates encased inside a glass vacuum tube, and cathode rays could be generated across the tube to the positive plate (anode) and detected when hitting a fluorescent screen or cross at the end of the tube. By using a magnetic field around the tube, the cathode rays could be deflected from a linear path, and due to the direction of the deflection, identify themselves as negative charges that we call electrons. We will see that once the magnitude of the charge on these electrons was determined by Millikan, we can also calculate the mass of the electron using fundamental Newtonian mechanics. *Every single electronic device we use today depends on our knowing the exact charge and mass of an electron.* With Planck's discovery that light energy is quantized in the emission spectrum of blackbody radiation, Niels Bohr surmised that perhaps the electrons in orbit around the nucleus of an atom also have quantized energy levels. He turned his attention to the emission lines in the hydrogen spectrum, which we have discussed previously, and which had been observed for decades but never fully explained.

The hydrogen atom was then visualized as a single proton nucleus with an orbiting electron going around it. Bohr made the assumption of quantized energy levels for this electron, meaning it was only allowed discrete energy values, and with some fairly basic analysis calculated what those energy values should be. From there, it was hypothesized that when the electron "jumps" energy levels, the difference in energy thus produced must be accounted for by the photons associated with the spectral lines seen under high voltage. Since only specific energy values were allowed, the photon energy emitted if the electron jumped down toward its ground state (lowest available energy) had a specific value that corresponds to a specific frequency and wavelength (recall $E = hf$). In this manner, Bohr was able to predict the wavelengths seen in the Balmer series of the hydrogen spectrum, and it worked! While this was a great success, it was not a perfect match with experimental observations of the hydrogen spectrum. There were some troubling issues with the analysis. First, the spectral lines, upon closer examination, were actually split into a series of lines, a phenomenon called fine structure. Second, the Bohr model was not successful in explaining the spectral lines of higher elements having more than one electron. Third, the Bohr model did not explain why the relative intensity of the spectral lines was different: some lines were brighter than others. It would take Schrodinger's (probability) wave equation, Heisenberg's Uncertainty Principle, and much further analysis over the next decades to account for these other observations. Nonetheless, it was a superb start to analyzing the atom and its structure.

Momentous discoveries were made in the decades leading up to World War II, at least one of which had a crucial impact on the outcome of the war, and which came within a whisker of being utilized by Hitler's brutal fascist regime. In the early 1930s, James Chadwick, using the conservation laws which we will explore in the concluding section of this book, discovered the neutron: the neutral particle of no charge and nearly the same mass as the proton that also occupies the nucleus of every element except hydrogen. (Isotopes of hydrogen contain neutrons, such as deuterium which is comprised of one proton and neutron, and tritium, which contains two neutrons and one proton. Isotopes are different nuclear versions of the same element.) After this discovery, the neutron was used to "probe" the nucleus of higher elements. Because of its neutral charge, unlike the alpha particles used by Rutherford, neutrons could penetrate far into the nucleus, and if that nucleus was unstable, as many of the larger elements tend to be due to the large repulsive forces of many protons crowded into that nucleus, the neutrons could cause fission to occur. It wasn't long after this that physicists split the nucleus of atoms such as Uranium-235 into smaller elements, albeit unknowingly (thankfully) in some cases, thereby creating the possibility of liberating huge amounts of energy. The historical events which followed are well known at this juncture, but it is worth noting that if all of these ideas being discussed were false, the atomic bomb would never have worked.

Yet another "particle" discovery was made in the 1930s, based on Paul A. M. Dirac's theoretical prediction that there should be two possible solutions to the energy states of the electron, simplified here by analogy. The equation $x^2 = 16$ has two possible roots; either +4 or -4 will satisfy this equation when substituted for the variable "x". When looking at the energy equations for the electron, Dirac suggested that since it was in the same form as the above example, perhaps *both* roots of the equation have physical meaning, and hence there might be an anti-particle just like the electron with the same mass and spin (another characteristic of particles we will discuss in the conservation laws section) but with positive charge of exactly the same magnitude as the electron. This particle was called the positron, and a few years after its prediction, Carl Anderson detected its existence in cosmic rays as they passed through his experimental cloud chamber, leaving a visible trail that looked just like electrons, except the path curved the opposite way in a magnetic field. It was soon concluded that *every* particle in nature has an anti-particle opposite, and it was also shown experimentally that when matter and anti-matter meet, the result is mutual annihilation with the mass of the particles transformed into the energy (a manifestation of the equation $E = mc^2$) typically in the form of gamma rays (high energy photons).

If this sounds like material right out of Star Trek, that is understandable, but antimatter-matter annihilation is one tool we use to do PET scans to detect tumors and other medical issues, so it is very real. PET scans stand for Positron Emission Tomography, which of course employs antimatter through the use of certain elements with unstable nuclei that decay or change to more stable forms by emitting positrons. In a very short time, usually nanoseconds, these positrons annihilate with "ordinary" electrons to form gamma ray photons which can then be detected in such a way as to discern variations in tissue structure that indicate abnormalities such as cancer cells.

Questions regarding antimatter abound. Why is there so little of it in our universe? Physicists love symmetry, so it seemed reasonable to propose there should be equal amounts of matter and antimatter in the universe. Obviously, there is far more ordinary matter than their rare antimatter counterparts, and that is a very good thing otherwise the entire universe would annihilate in a massive shower of gamma rays and we would never have lived as a bunch of organized atoms, pondering our existence! Something quite fortunate must have happened in the first moments of the Big Bang to create the asymmetry we observe today. We are still trying to figure out the process or processes that may have occurred to produce what we see all around us. Feynman and others have proposed that perhaps positrons are actually electrons traveling backward in time. Though this approach yields some correct results experimentally, it is not clear what it means physically. Are there galaxies composed of mainly antimatter and a tiny slice of ordinary matter? Though doubtful, it is not impossible; moreover, perhaps there are parallel *universe*s made of antimatter. Speculation along such avenues is fun and, at times, can lead to amazing breakthroughs in our understanding of the universe we live in, but it is also important to distinguish such guesswork from experimental verification. For now, we simply do not know.

The mysteries inherent in mass deepen dramatically when we study spiral galaxies, which occur in the shape of a squashed disk. In such galaxies, all of the mass rotates about the galactic center due to mutual gravitational attraction. Think of running down a hallway, and then some mischievous person lassos you and, because of your inertia, you move around him in "orbit". The harder he pulls on you, the faster you must run to maintain that orbit. The same mechanism applies to matter as it moves around the center of a galaxy, and we shall see when we study the force of gravity that if we can figure out the speed of all this mass in orbit, we can deduce the entire mass of the galactic core - this is one way we can estimate the number of stars in the Milky Way or any galaxy! The other way to estimate the mass of a galaxy is to measure its brightness due to the combined luminosity of all of its stars. In theory, these numbers should closely match one another, since there is only one correct value for the mass of anything, plus or minus a small uncertainty in measurement. We measure the speed of the rotating mass in a galaxy by going back to the Doppler Effect: the leading edge of the matter moving towards us should be Doppler shifted to higher (blue) frequencies, and the tailing edge should be red-shifted due to its recession from us. These shifts are separate from the Doppler red shift due to the universe's expansion discussed previously. By measuring the amount of shift, we can deduce the speed of the rotating mass. Many such measurements have been made, looking at a large variety of such galaxies. *All the experimental results indicate that the total mass of galaxies found from looking at orbital speeds is almost five times greater than the mass obtained by looking at the combined brightness of all of its stars.* The inescapable conclusion appears to be that all the ordinary matter (and even antimatter) that we know about (trees, mountains, people, stars, oceans, and so forth) is only a tiny fraction of what mass there actually seems to be in the universe! The much more prevalent mass in these galaxies remains as an unknown; we call it "dark matter" because we certainly can't see it, yet it must be there.

    A still greater mystery is the *acceleration* of the expanding universe we discussed previously, an experimental fact we have no explanation for at this time. We call it "dark energy", which of course explains nothing. According to the data, the amount of matter in the universe breaks down as follows: about 70% must be dark energy, around 25% must be dark matter, and the remaining 5% is the ordinary matter (atoms and such) that make up you and me and everything we have observed on our planet and in our universe. The bottom line is a terrific lesson in humility: as much as we have progressed, we still have no clue as to what constitutes *95%* of the universe's mass.

As we did for space and time, we need to define a standard for the concept of mass that everyone can agree on and use as a prototype for all applications. First, we must be sure not to confuse mass with size or volume. While it is obvious that a spherical chunk of iron has more mass than a large hollow beach ball, it is nevertheless easy to fall into the trap of thinking larger sized objects must have greater mass. Likewise, we must not confuse mass with weight, even though the terms are often used interchangeably. Weight depends on location, mass does not. A trip to the Moon demonstrates this fact dramatically, as a 180-pound person would weigh about 30 pounds on the Moon's surface, but the amount of matter or atoms making up the person does not change (significantly) due to the trip. Prospectors mining gold could make a fortune by buying precious metals at high elevations where gravity is weaker, and then selling it at sea level where it weighs more, if we chose to measure things out that way. Instead, using equal arm balances, we can compare equivalent masses to a precision of nearly one part per billion. Clearly mass is directly proportional to weight, since more mass leads to more weight, but that does not imply they are the same.

Perhaps we could measure mass by the amount of resistance it offers to a change in state of its motion. This is tantamount to measuring the concept of inertia, and necessarily involves introducing another physical quantity: force. Again, while mass is directly related to inertia, to equate the two terms is not useful in terms of defining a standard. Density is defined as the amount of mass per unit volume, so perhaps that could be the avenue of approach. Water is readily available to everyone, hopefully, so by using the fact that water is most dense at 4 degrees Celsius, we could develop a standard for mass from that knowledge. Incidentally, water is a most extraordinary substance, one that *expands* when cooled down until it forms a crystalline structure (ice) that now occupies a greater volume and thus actually reduces in density even though the amount of mass has not changed. Nature again shows her remarkable plan in all of this, for if water was like nearly all other substances, it would be denser in the solid state, and the ice would sink to the bottom of rivers, lakes, and oceans, killing all the fish. The problem with density as a measure, even though it can be quite useful to identify substances since every substance possesses a (somewhat) unique density, is its variation with temperature.

Defining a standard mass, the kilogram, turns out historically to be the most difficult nut to crack of them all. Though one cubic centimeter of water at 4 degrees Celsius was considered as the standard for one gram of mass, eventually it was decided to make one kilogram a cylindrical block of platinum-iridium and use that as the prototype for all mass measurements. This cylinder was stored in Sevres, near Paris, much the same as the standard meter bar made from the same alloy. This "international prototype kilogram", as it was called, was used as the standard for all reproductions of the kilogram throughout the world for many years, right up to May 20, 2019. The new standard now employed by international agreement is based on an exact measurement of the value for Planck's constant h, using a prescribed experimental method: $h = 6.62607015 \times 10^{-34}$ Joule·seconds. A Joule is a unit of energy in the metric system, as we shall see further on down the road, but in basic units one Joule equals one kilogram x meters squared divided by seconds squared. In the final analysis, the kilogram is predicated on measuring the speed of light (used to measure the space called one meter) and the vibrations of the cesium atom (used to measure the span of time called one second). By knowing how to measure "h" with a high degree of precision, one can work backward to develop the amount of matter that makes up one kilogram with better precision than counting atoms, measuring out water, or depending on a metal cylinder that could be stolen or lost. We now have a standard for each of the three concepts of space, time, and mass, which together encompass a huge swath of physics.

Before leaving this concept of mass and voyaging into the fourth and final basic concept, charge, a couple of loose ends need to be addressed. Mass is a scalar quantity, not a vector. It carries with it no information about direction, only magnitude. For example, we would never say or write that the mass of a car is 2000 kilograms, west. When we use mass in physics, one number is all that is required, and it does not involve any direction. The question arises that since space and time have proven to *not* be absolute, or the same to all observers, is the same true for mass? Experimental results show, repeatedly, that indeed mass *does* change depending on one's reference frame. Specifically, the mass of any object increases relative to a stationary observer as the object approaches the speed of light. We have observed this phenomenon many millions of times over the years, as our technology has developed to the point where we can accelerate electrons and protons (and other subatomic particles) close to the speed of light in huge machines, such as are located at CERN and several other locations around the globe. All of these billion dollar machines would be useless if mass expansion was not taken into account, by exactly the amount Einstein predicted in his Special Theory of Relativity.

Since the speed of light is finite, and one cannot go faster than that speed in our universe as far as we know, adding energy to a moving system transforms into a measurement of greater mass rather than greater speed. This suggests energy-mass equivalence once again, and brings to mind that most famous equation which most people know and is so often misunderstood: $E = mc^2$. We will arrive back at that famous equation at the end of this book.

We are left with incredibly precise standards of measure for space, time, and mass, but also with the realization that *all three concepts are relative and not absolute*. This is a most extraordinary universe we live in, one that is filled with common-sense contradictions, unexpected nuance, wild beauty, and wonderful complexity. Onward to the concept of charge!

CHARGE

We've arrived at the final fundamental concept of the four under discussion. Thus far we have witnessed how pervasively these basic ideas labeled space, time, and mass weave their way into the field of physics: they reach down to the very roots of the discipline and spread through all the various branches in thousands of "capillaries", all of which are the life-blood of this work in progress to understand nature and her patterns. This fourth concept we have known about for thousands of years, at least in a qualitative way. When certain materials are rubbed together, such as wool and rubber or silk and glass, they acquire an ability to attract pieces of paper or straw. Rub a comb through your hair and this property of matter becomes readily observable when the comb is placed next to a steadily dripping water spout: the water curls in toward the comb. Rub a balloon on your hair and stick it to a wall!

In one of my favorite such demonstrations, place a long meter stick or 2x4 wooden stud on a watch glass or some type of fulcrum so that it is balanced and free to rotate. Now take a rubber rod (usually ebonite) and a wool cloth, rub the two together vigorously for half a minute or so, and go to either end of the balanced meter stick, placing the rod close to but not touching the stick. When you do this, especially on a dry day, it will rotate toward the rubber rod in what looks like a magic trick! Since all such items are made from atoms, and atoms have positive and negative charges, perhaps it is not surprising that these processes are so readily observed. Yet as children, we all seem to share an inherent wonder and excitement regarding the universe we experience around us, and we tend to ask fantastic questions about it all. If we are fortunate and work hard at it, some of us retain that sense of wonder for a lifetime. For too many of us, that initial awe we experience becomes obscured by the ordinary demands of adult life, and sometimes (unfortunately) by having it crammed down our throats far too often as "things to memorize". As a child, the immediate observation is all that matters; it is an awareness which requires no analysis or rationale. As the years pass we learn what others thought about and reasoned when confronted with these phenomena, and this can be extremely helpful, as long as that process does not result in closing down all other avenues of approach, *especially your own*!

It has been proven, many times over in the history of science, that sometimes an idea comes along which takes our understanding to a whole new level, and it almost always is an idea that is a completely new way of thinking. There is a balancing act to be made here, between listening and studying the work of others while simultaneously giving your own thoughts and ideas a firm foothold. Do not let the "experts" coerce you, subliminally or otherwise, that your ideas are not worth much. Indeed, the ideas may be completely wrong, but that is separate from claiming they have no worth. Often, we find it is our mistakes and false dead-ends that direct us to more fruitful paths. When we immerse ourselves in nature again, as adults, we are then ready to meet the world face to face once more, this time with an open, free, and curious mind that will be amazed in ways that may go much deeper than those of a child. That is the mindset I hope we achieve as we climb up this fourth (metaphorical) mountain called charge.

In ancient Greece, the yellow resin we call amber, found in evergreen trees, was given the name "elektron". They discovered that when this amber was rubbed with certain materials such as cloth, a crackling sound was often heard and bits of straw would fly up as though they were somehow attracted to the amber. By the 19th century, particularly with so much experimentation being done within the relatively new field of electricity, the term used to identify electric charge was formally named the "electron" by an Irish physicist George Johnstone Stoney, in 1891, and clearly with a nod to the etymology of the word. Charles Augustine de Coulomb came up with a quantitative formula for the force of attraction or repulsion between two stationary electric charges. This was done about the year 1800, and some facts had begun to emerge about this concept of charge:

1. Electric charges come in two varieties only, positive and negative.

2. Like charges repel each other, opposite charges attract each other.

3. The force between two electric charges increases when the amount of charge increases.

4. The force between two charges varies as the inverse square of the distance between them.

This last item (#4) may sound a bit convoluted, but it recurs throughout nature in other forms so it is important to understand what it means. Two examples may serve to illuminate how this discovery is applied. Suppose the force between two charges is measured to be 64, in arbitrary units of force (the next section on the fundamental forces will address units and much more). We then double the separation distance between the charges and ask: what is the electrical force between them now? Since we doubled the distance between them and the inverse of that is squared, we are reducing the force by $(1/2)^2$, so the force is reduced by a factor of 4 and now equals 16. As the second example, and using the same initial force of 64, assume we halve the distance between the charges, bringing them much closer to each other. One would expect that the force would increase, at least that seems intuitive. Indeed, that is what happens. Half the distance implies four times the force, or now 256 units. That is how all inverse square laws work. Identify the factor involved, take the reciprocal (inverse) of it, then square it. The number you are left with, multiplied by the original force, is the new force!

There have been some hard lessons to learn in this particular climb called charge. It took some tragic accidents in refueling jets to understand that the high velocity of the jet fuel passing through a rubber hose could generate static charge, and the spark can set off the high-octane fuel. You may have noticed warnings about this danger when at the gasoline station: turn off the cell phone, don't re-enter the vehicle and return during refueling, especially during the dry months of winter. In the fueling process for aircraft, we now provide a safe path to the Earth, called grounding, for the buildup of charge to safely find its way to the ground without the potentially fatal static spark. It is suspected that it was static charge buildup that caused the spark which ignited the Hindenburg (which was already extremely dangerous since the craft used highly flammable hydrogen gas, rather than the inert helium that is now used in blimps). Trucks carrying flammable liquids are now manufactured with special tires that conduct static charge into the road as they travel, or they may use grounding straps or chains to safely discharge the vehicle. There are many such examples of the danger of static charge buildup, most of which go far beyond the annoyance of the static cling found in dryers and clothes. The granddaddy of all of them though has to be the one everyone has had experience with, hopefully at a safe distance: lightning.

 Though Benjamin Franklin is perhaps the most famous person to investigate the supposed connection between lightning and electricity, he certainly wasn't the first or the only one to do so. Most historians agree that he actually did fly a kite in a thunderstorm, something I would definitely not recommend to anyone, but it can't be completely confirmed since he did the experiment with only his son. On balance, I would wager it occurred as he said it did. With hemp string, silk string, a key, and a device called a Leyden Jar that stores electric charge and was a precursor of what is now called a capacitor, Franklin flew the kite in a gathering thunderstorm in 1752. It is almost certainly *not* true that the kite was hit by lightning, since if it had been, the probability of him and his son surviving would have been extremely low. Far more probable is that the charge accumulating from the storm clouds found their way down the hemp line and arced over to Franklin when he stretched his hand out to touch the key. The upshot of the experiment was clear evidence that lightning was, in fact, an electrical phenomenon. He went on to invent the lightning rod, a pointed metal object with the purpose of *attracting* lightning away from buildings and people and conducting the excess charge and energy safely into the ground. Obviously, a poorly installed lightning rod is far worse than none at all! It is fortunate for the U.S.A. that he survived the dangerous kite experiment, for that allowed him to play a pivotal role in the genesis of a new country founded on aspirations of freedom, even though the application of those freedoms did not extend to large segments of the population.

It has taken many years and many thousands of studies to understand the mechanisms at work when lightning occurs, and there is still much we do not fully know. Through what might be termed a frictional process, something like rubbing one's feet on a thick carpet, one hypothesis is that the constant motion of water, ice, air, and other particulate matter circulating in a cumulonimbus cloud can create a huge excess of negative charge at the bottom of the cloud, though there are pockets of exposed positive charge within the cloud as well. This in turn, since like charges repel, forces the negative charges in the terrain below to move away (downward) from the bottom of this cloud, exposing a net positive charge at the surface, especially at the highest, sharpest points of the terrain. This is a process called polarization: the separation of charges in a closed system. Though dry air is an excellent insulator, which is why we fill down parkas with it, water moisture, dust, pollen, and things of that ilk are much better conductors of charge, and natural processes do not tend to sustain this type of huge charge separation for any length of time. In brief, typically several paths are provided for the opposite charges to neutralize one another, and nature will find these paths given the time to do so. Let us slow down the process and look at what we have found actually happens in a lightning strike. Again, this is an ongoing field of research!

When paths are established between the huge negative charge stored in the belly of the storm cloud and locations within the cloud that are positively charged, channels are formed for the charges to neutralize. In the process that is perhaps best understood, ground-cloud lightning, often these channels are not perfectly balanced, and excess negative charge "leaks out" of the cloud and follows the paths down close to the ground using those water droplets, dust particles, pollen, and so forth as the means of travel. These are called the "dark leaders", and we do *not* see them in most cases. When the negative charges come close to the exposed positive charge in the terrain, channels of connection occur and many of the dark leaders light up in the return main stroke, which is the lightning we see. This is a truly massive release of energy, carrying with it about five times the surface temperature of the sun, and a rapid movement of charge in a tiny fraction of a second. These huge temperatures cause the air to expand rapidly and then cool, which is the cause of the shock wave and thunder heard after the bolt strikes.

Contrary to popular belief, this does mean that almost all lightning bolts that we see between the clouds and ground consist of the main stroke actually traveling upward from the ground to the cloud, although it happens so rapidly that only a high-speed time exposure can capture it frame by frame. This also indicates that in a very real sense, this is the same static discharge one experiences on a dry, winter day when sliding across a car seat, just on a far greater scale! There are several types of lightning: heat lightning, cloud-cloud lightning, ball lightning, and so forth., but cloud-ground lightning is perhaps the best understood since observations for these strikes are more readily available. I have greatly simplified the process here, and I invite the reader to investigate this spectacular phenomenon further. I can only imagine, again, what the first humans were feeling and thinking when they experienced their first violent thunderstorm. There are several myths surrounding how to stay safe when lightning is in the area, and we should explore those myths before moving on.

We picture stationary electric charge as producing an electric field (E) around the charge, much the same as mass produces a gravitational field around itself. Note that the electric field is a vector, because direction as well as magnitude is important to specify. Electric field lines, by convention, go into negative charges and out of positive charges. The closer the field lines are to one another, the stronger the field. Picture rays of sunlight streaming out from the surface of the sun – this is how the electric field surrounding a positive charge can be visualized. This concept of field was invented by Michael Faraday, a tremendously imaginative British physicist in the 19th century. He was an outstanding experimentalist and visualized nature in profound and productive ways; we will meet him again shortly. The first myth we need to dispel is that the rubber tires of a car are what keeps us safe in a thunderstorm. This is false; for rubber to insulate us from the amount of energy released in a lightning bolt, it would need to be miles thick. *It is the metal frame of the car which keeps us from harm, because the electric field inside any hollow conductor, such as a metal, is zero.* Perhaps you have seen this in a field trip to a museum of science, where a brave volunteer stands inside a large metal cage that looks like a birdcage and gets "zapped" by huge static bolts produced by Van de Graff generators. The discharge stays on the metal surface of the cage or car, and the net electric field inside the cage adds to zero, keeping the volunteer (and you in a car) safe. Though an automobile is not a perfectly closed system, the probability of lightning finding its way inside a car (windows up, not touching metal surfaces) is exceedingly small.

Other myths deal with how far away one can safely be from the first sighting of a lightning bolt or the first sound of thunder. Here's the scoop: if you see lightning or hear thunder and you are in an open area, stop counting and get moving. The general rule for *all of electricity* is to never become the connecting link between two different "potentials", which we will come to define as voltages. Nature will take the paths available to neutralize a separation of charge or to move from an area of higher energy to lower energy: do not get in her way! Therefore, *avoid at all costs being the highest object along the terrain you are in during a storm, and equally as important, avoid being near the highest object.* A lone, tall tree may be great shelter from the rain, but it is also the perfect "widow maker" and the place most people who are killed by lightning seek refuge. It is true that it is best to stay off the phone, stay out of the shower, stay away from windows in a violent storm, for many reasons. Lightning has been known to travel along phone lines, through plumbing, along the ground through tree roots, and basically through any conductive path it can find. The metal cage is your best bet, because the loosely bound electrons in metals make for excellent conductors, and nature will take the path of less resistance. If you are outdoors, seek shelter immediately inside a building, and if you can't find a building nearby, make sure you are not the tallest object or near the tallest object in your surroundings. Lightning need not be feared, but it most definitely needs to be respected.

We have been looking at the field of physics called electrostatics, which is the study of electrical processes wherein electric charges are stationary and remain stationary in most cases. The discussion now moves to electrodynamics: moving charges. This includes the familiar electricity we speak of when talking about plugging in appliances, lighting lamps, and so forth. Since we have discovered that the three other basic concepts in physics called space, time, and mass have turned out to be relative and not absolute, it is reasonable to ask if moving charges have the same property of relativity. Namely, does the amount of charge vary with speed (or any other factor) as measured by a stationary observer? Surprisingly perhaps, the answer this time is no! Charge, according to all the experiments we have done and all the evidence we have, does not change its amount due to relative motion: we say it is *invariant*. Moving charge does, however, produce some amazing phenomena that we will get to in a moment.

Electric current usually involves moving charge in metal conductors, and it is measured in Amperes, in honor of Andre-Marie Ampere, a French physicist who was a pioneer in the development of electrodynamics in the early stages of the 19th century. Staying with our emphasis on the four basic concepts, one Ampere, typically truncated to one Amp, is equal to charge/time. We measure charge in terms of Coulombs, so 1 Amp = 1 Coulomb/second. Keep in mind as we progress into electromagnetism that all the units used, many of which are the names of physicists who did exceptional work in the field, always reduce back to some combination of the four basic concepts under investigation. There is considerable confusion regarding electric current, as people often visualize it as similar to water flowing through a pipe, blood through arteries, or some mass movement of charge through a wire. While these analogies have some benefit in terms of learning the physics involved, it is important to understand that the loosely bound electrons in metal conductors such as wires do not travel from the switch to the filament of the light bulb! If that were the case, it would take years for the charges to reach the filament and cause the light to go on. It is reminiscent of the radio disc-jockey who played a practical joke on April 1 one year, convincing people to take apart their phones to empty out the "electron dust" that supposedly accumulates in the phone over time. More accurately, electric current is like a breeze blowing across a wheat field with the strands of wheat (electrons) bending in the direction of lower energy. The electrons drift or oscillate, thus becoming conveyors of the *energy* that is being transmitted close to the speed of light.

Voltage, a term in honor of Volta who invented the battery around 1800, along with Galvani and perhaps others, is the amount of energy per unit charge in the circuit. If we carry the water through the pipe analogy forward, we think of voltage as the pump or source of energy, and the water flowing as the current. Again, this is not entirely accurate, but the comparison often helps people distinguish between voltage and current. As we will see in great detail, the unit of energy is a combination of mass, space, and time, and in the metric system is given the name Joules in honor of James Joule who will visit us later. Therefore, voltage also breaks down to our four fundamental concepts. Another misconception when dealing with electricity, one reinforced by signage one can see near power lines and stations, is that high voltage kills. It can, and often does, but it is not the high voltage that is lethal, it is the current which flows when one connects a high voltage to a low voltage (often 0 volts or "ground" – literally the Earth). We can prove this by observing birds perched on extremely high voltage power lines: they chirp away happily without any detrimental effects up there! Should they accidentally brush against a line or a surface of much lower voltage, they then become the connecting link between two different voltages and the current that ensues will be deadly. It doesn't take much current to cause major issues in the human body: .02 Amps can cause difficulty in breathing, and up to .10 Amps or more can cause the heart to fibrillate. Hence the admonition once again: *do not become the link between two different voltages.*

There are two types of current, each with its advantages and disadvantages, and each generated in specific ways. Direct current, abbreviated DC, is the type of current that is produced by a battery typically. This type of current moves in one direction only, and by convention (largely due to our friend Benjamin Franklin) we take the direction to be from the positive terminal toward the negative terminal. The actual movement of the electron drift is reversed, but nature doesn't care about our sign conventions as long as we remain consistent. The other type of current is alternating current, or AC, and this type is generally deadlier than DC because it involves vibrating charges back and forth through time in a periodic "dance" that transfers energy. In the United States, this alternating frequency is 60 cycles per second, so if one comes in contact with an AC power line, every cell in the body which runs off electricity starts to do the 60 cycle/second jig, and that includes breathing and the heart. Muscles, including the heart, like a frequency much lower than 60 cycles/second, so either they freeze up or spasm with AC electricity, and that is not healthy of course. The generation of commercially available AC lagged several decades behind battery technology, and led to the epic battle between Edison and Westinghouse which we shall see.

With the invention of the battery, new technology once again assumed the role of driving progress in science forward, which in turn drove still newer technology forward as discoveries unfolded on a regular basis. If communication systems had been as fast in the early part of the 19th century as they are now, the pace of advances would have been far greater. It often took months, sometimes years, for new discoveries and theories to be shared across the continents. During this time, Georg Ohm investigated the relationship between voltage and current in a circuit built with copper wires and batteries. He discovered that for most such circuits, the ratio of voltage/current remained constant! While it is not surprising that more voltage at the source would generate more current, the fact that the relationship is linear for metallic conductors of this nature was neither known nor apparent. The ratio was dubbed "resistance", and of course given the unit called ohms. This became, at least initially, the foundational principle guiding the study of electrical circuits, and is called Ohm's Law. The law states that voltage = current multiplied by resistance, or written in shorthand: V = IR, where V = voltage, I = current, and R = resistance. One can see that ohms, given the symbol omega (Ω) from the Greek alphabet so as not to confuse it with a zero, also reduced to the four concepts since voltage/current each do as well. As one example, a 12-volt car battery hooked up to a circuit having a resistance of 6 Ohms (6 Ω) would cause a current of 2 Amps in the circuit.

The inquisitive reader might look at that example and wonder why a person touching both posts of a car battery (***never do this while someone else is cranking the engine or starting the car***) may not even feel a mild shock. The answer relies on the fact that in dry, normal conditions, the internal resistance of a human being is typically on the order of hundreds of thousands of ohms, so the amount of current is probably going to be too small to detect. But to reiterate, when the engine is cranked, the voltages produced by other electrical components in the car is much higher than 12 volts, and touching the posts of the battery is risky business. Another safety tip to keep in mind is that water is a polar molecule, with the hydrogen side of the molecule being positive and the oxygen side negative. This makes it responsive to electrical systems, which is why you can bend a stream of water toward you if you place a charged rod near the stream. Water also can have the effect of drastically reducing your internal resistance, making even small voltage differences dangerous. Hence, water and electricity do not mix well, and to avoid a bad shock or worse, make sure you do not become another statistic by testing that hypothesis.

How does one make a resistor? It's an interesting engineering question, and the simple answer is that *anything* that has electrical energy running through it has a resistance, by default. Therefore, lightbulbs, toasters, refrigerators, wires, switches, and so on each has some resistance associated with it that depends on a number of physical factors. I am betting that the reader can guess at what some of these factors might be, starting with the material the resistor is made from: clearly copper wire is going to have less inherent resistance than a strip of rubber. We call this a measure of resistivity, and each material has a value that one can look up in a table because some very diligent folks have measured it for a living! Another factor is the thickness or cross-sectional area of the device, as it would make sense the wider (greater) the opening for the current to flow, the smaller the resistance will be, an inverse relationship. This is why the wires to your clothes dryer and oven are much thicker than the other wires in the house, because each requires more current to operate, so each must have less resistance in order for the wires not to overheat and cause a fire hazard. The third factor is the length of the resistor: the longer it is, the more resistance, a direct relationship. The combination of these three factors is what determines the resistance of those color-coded cylinders one can see inside an electrical device.

When I think of resistance, the analogy that comes to mind often is people trying to exit a theater after a movie. If the exit is a wide-open door, then the "current", the people leaving, is large. If the doorway is narrow and involves a long hallway, the current reduces to a trickle. This analogy comes in handy when discussing the two types of circuits most often analyzed and used: series and parallel. In the spirit of Columbo, that somewhat irritating but extremely efficient fictional detective, just one more question before finishing up with circuits for a time: what effect does temperature have on resistance? And as a corollary: what does temperature measure? Tackling the latter question first, in essence temperature is a measure of molecular motion. The hotter something is, the faster the molecules jiggle about on average. This is the primary reason most materials expand when heated, since the greater speeds result in occupying a larger volume. One can then see that hot air is less dense because it occupies a larger volume, and that is why the convection process results in warm air rising and expanding while the cooler air settles below. Substances in the solid state have only the vibrational mode of freedom: the atoms or molecules vibrate about a fixed position. As more energy is supplied to the solid, eventually the molecules gain more freedom and are able to rotate as a liquid, which is how liquids can pour and assume the shape of their container. Still more energy added results in the molecules gaining complete freedom, and they are able to translate, rotate, and vibrate through space as a gas. If we continue to heat a gas for a long period of time, we might also attain a plasma: ionized gas whose atoms are stripped of electrons, becoming the stuff that stars are made from in order to undergo the fusion process.

One might reasonably ask: what is the hottest temperature we have encountered, and is there an upper limit to temperature? The machines which smash particles together, such as CERN, have generated temperatures in the trillions of degrees here on Earth, and that is a quarter of a million times hotter than the center of the Sun. Still, it is a paltry number in comparison to what is called the Planck temperature at the beginning of the universe, estimated to be around 10^{32} degrees Kelvin, and above which the laws of physics as we know them break down completely. As far as an absolute upper limit, nobody knows what that might be or if the question even has an answer.

Moving in the other direction and connecting back to electrical resistance, what happens when we cool certain materials down? Predictably, most materials experience a steady drop in resistance along with a corresponding drop in temperature. But then, for a few known substances, a specific temperature is reached where *suddenly the resistance abruptly drops to exactly zero!* This is a most extraordinary phenomenon, discovered in the early decades of the 20th century, and one fraught with all sorts of possible applications. Imagine if computers could run with zero resistance – they would leave even the fastest processors in the dust. Of course, this is the process called superconductivity, and it is very much like a phase transition. As one example, mercury steadily loses resistance as the temperature is reduced, and then suddenly at 4.2 degrees Kelvin and below it has exactly zero resistance. At some point, it appears the electrons suddenly communicate with one another in a way analogous to people exiting that theater by creating a channel for everyone to flow out unimpeded. A superconducting current set up in a wire would still be there, intact, millions of years from its inception.

Helium, typically a gas at normal temperatures and pressures, transitions into a superfluid at 2.17 degrees Kelvin, and at that point loses all resistance to flow (zero viscosity). This makes for some wild observations: liquid helium pouring through the *bottom* of a solid container, or shooting up the sides of the container, seemingly with a mind of its own. If we set liquid helium into a rotating swirling eddy and figured out a way to revisit the scene millions of years later, it would still be swirling about unaffected by time! These temperatures are extremely low, close to the mark of minus 460 degrees Fahrenheit, which is defined as absolute zero degrees Kelvin, or minus 273.15 degrees Celsius. The closest we have come to absolute zero experimentally is now in the vicinity of tiny thousandths of a degree above 0 Kelvin. We will see that to actually reach absolute zero carries with it a whole slew of problems, theoretical and otherwise, and so it may not be possible to ever reach that exact point. Research is ongoing, and the practical applications of superconductivity continue to be explored by developing ceramic materials that can become superconducting at much higher temperatures, with the latest at this writing being about 250 degrees Kelvin.

Some materials are good conductors, such as metals, because their outer shell (called valence) electrons are loosely bound to the nucleus and are thus available to be used for energy transport of many different types. Other substances are good insulators, with tightly bound electrons that are quite content to stay right where they are in their current state. Then there are the interesting elements that are halfway between, called semiconductors. Carbon is one example, having four electrons on its outer shell. We can build the periodic table, and consequently the electronic structure of each element, from some basic rules in quantum chemistry. When we return to the atom we can investigate this further, but for now we will use the fact that this particular outer shell has the most stability when it contains a total of eight electrons. Nature tends toward stability, so carbon atoms form long chains, each sharing their four electrons with one another (to make the eight) in a covalent bonding scheme that drives most organic chemistry students nuts. This ability to form long chains of molecules is also a main ingredient in the production of life forms based on carbon, like you and me. Carbon isn't the only element that has the four outer shell electrons – silicon is another example, and so we can build a platform whereby chains of silicon atoms can co-exist. It is called a semiconductor because the electrons in this outer shell are not quite available for energy transport, so we say they reside just below the "conduction band", and a relatively small impetus can push these electrons into the conduction band.

Now suppose we get the idea of "doping" this silicon wafer with small amounts of impurities, and see what happens. The impurities have to be chosen judiciously and at very low concentrations compared to the silicon chains. We might use boron, with three valence electrons, and phosphorus having five valance electrons, and situate them in clever ways within the silicon. We can think of an analogy where there are eight parking spaces available, and now we have a situation where we can move the "extra" fifth electron into a silicon atom, which then has the extra electron that will fill the "hole" where the boron exists, so that all the spaces are filled up.

The beauty of this is that it can be done in very small spaces (getting smaller all the time), and only a small voltage is required to move that "extra" electron into a hole. We have developed a sophisticated switch, so that when we introduce the small voltage, we can code this as a binary "1", and with the switch off we have a binary "0". Each key we hit on the keyboard of a computer is then processed in a complex web of binary logic, using the basics of semiconductor electron movement to guide the whole enterprise. Welcome to the fantastic world of computers!

We will return to the topic of electrical circuits in the last section of the book, and introduce a methodology that will solve *any circuit*, no matter how complicated. It turns out that Ohm's Law has several limitations; in spite of its usefulness in many common circuits, when applied to a variety of electronic circuits, its validity breaks down. For now, let us look at two types of wiring employed in circuitry that were mentioned in passing a while ago: series and parallel. In series wiring, all electrical components are strung together, one after the other, so that every component gets the same current going through it. Therefore, just like those old strings of Christmas tree lights, if one element fails then the entire circuit fails. Also, since there is only so much energy available in the circuit, by connecting them in series the source voltage has to be split among all the elements, with more voltage required where there is more resistance. One can deduce that the total resistance of this type of circuit is the sum of all the individual resistors, analogous to a number of consecutive doors leading to the outside world from a theater. The "people" current must pass through each door to get out, and each door requires energy to get through, so the total current is going to drop as the total resistance gets bigger. Moreover, if one door is locked, the current stops completely because there is no other way out.

Parallel circuits offer a much different scenario, and as you will see, this is how the wiring in your house is done. In parallel, each circuit element receives the source voltage independently from all the other elements. Therefore, in a house circuit for example, all the lights, appliances, and so forth on one parallel branch are connected across the same source voltage, which in most cases in the United States is 120 volts. This means that each element "gets" the full 120 volts or close to it, and each element runs independently from the others. This offers distinct advantages: there is no dimming of the lights as one adds more load to the circuit, as there would be if the same was done in series, and if one element is off all the others can still be on. It would be a genuine nuisance if one had to turn on the toaster to make the kitchen lights work! This parallel branch is then connected independently into the circuit breaker box, or fuse box if the house is very old. We will return to breakers and fuses in a moment. The price one pays for the fact that each element receives the same source voltage is that more total current is being drawn to make that happen. This implies that as one adds elements in a parallel network, the total resistance goes *down*. Actually, using our analogy of people leaving a theater, this makes perfect sense. Imagine the theater now has multiple exit doors, and we keep adding more possible exits: the result is that more people can exit the theater at a much faster pace! This is tantamount to larger current resulting from lower resistance.

Let's tie up a few loose items before returning to fuses and circuit breakers. The actual voltage supplied to homes in the United States is around 240 volts. Three main wires enter the house: a +120 volt line, a neutral or ground line at 0 volts, and a -120 volt line. Each wire is color coded so electricians know which line they are dealing with. For the oven and clothes dryer, connections are made across the lines so that the full 240 volts is utilized; for all others, connections are made to obtain the 120 volts needed, since the specifications for those appliances are manufactured for that voltage. People who travel to Europe know that they need adaptors (transformers which step up the voltage) to have these appliances run properly, since Europe operates at a higher voltage setting. As an aside, the label "parallel" does not mean that the elements have to physically be parallel to one another, though that is often how this type of circuit diagram is drawn. All that is required for elements to be in parallel is that there is a junction between them for the current to split up and that each has an independent path to the source voltage.

Another concept that is used in physics is that of power. In general, power is defined as the amount of energy being used per unit time, or Joules per second. In turn, to honor the inventor of the steam engine, a Joule/second is called a Watt. You may also have heard of horsepower, which literally derives from the power of the "average" horse, and amounts to this horse pulling a 550-pound sled one foot in one second. This implies that a Porsche 918 Spyder has 875 horses pulling it along! For the sake of equivalency, it takes 746 Watts to make one horsepower. Regardless, keep in mind once again that whether it is Watts or horsepower, the units reduce to space, time, and mass. Electrical power is calculated by multiplying voltage (Joules/Coulomb) time current (Coulomb/second). Note that when this is done, the charge (Coulomb) unit cancels and we are left with Joules/second, a simple example of how dimensional analysis can be quite useful. Therefore, when the power company charges a set rate per kilowatt-hour, they are charging you for the electrical energy being used (since power x time = energy). For example, if the charge is 25 cents per kilowatt-hour, and one operates a 100-watt light bulb for 10 hours, that is 1000 watt-hours, or 1 kilowatt-hour, and you owe the utility company 25 cents. Now let us examine circuit breakers and fuses.

Each parallel network in your home has a total current rating, usually 15 or 20 Amps. Here is the critical piece of information: *the wiring in that circuit is specifically designed to handle that total amount of current safely without overheating and causing a fire.* If more appliances are added to the circuit through extension cords, or if a bunch of hairdryers were suddenly used on the same line for some reason, then the current drawn exceeds the rating and the circuit breaker trips to shut down the line completely and save your house from possibly burning down. This mechanism works off a simple principle we have already discussed: metals when heated expand and when cooled contract. The breaker is typically a bi-metallic strip, and it is made so that when the total current drawn exceeds the safe limit, the two strips expand, separate, and the circuit is shut off. This is how the old thermostats used to work, or sometimes mercury was used in a way to establish an on-off switch based on temperature changes, and now can be accomplished with sophisticated electronics. Once cooled after a short time, the breaker can be re-set, hopefully this time without the extra load being used again!

Fuses operate in the same way, but are far less convenient and offer the user some very risky options. When a fuse "blows", the high current causes the metal strip to literally melt, shutting down the line. People then try to outfox the safety feature, and may put a nail or penny back in place of the fuse. *This is extremely foolhardy and dangerous*, because now a huge amount of current can flow through the wires (it takes hundreds of amps to melt a penny), the wiring will catch fire, and if your house burns down it may be fatal and the insurance will not cover it because this is gross negligence. In the event of an electrical fire, such as a toaster that starts to flame up as an example, much like grease fires, *never throw water on it* as it will only make it far worse. Unplug the appliance from the source, use a fire extinguisher (one should always be handy), and call the fire department. If possible, suffocate the fire with a wet towel or blanket and bring the appliance outside. It would be remiss on my part not to include these facts, because physics is everywhere, and the laws of nature are to be respected. As I have alluded to many pages ago, every time we choose to remain ignorant of her laws, people suffer and often die.

Moving charge can occur through space *without* wires of course, and it is this phenomenon we will look at before closing the section on the four basic concepts in physics. In the late 19th century, advancing technology allowed J.J. Thomson and others to develop vacuum tubes with electrodes embedded inside. When high voltage was applied to these electrodes, it was observed that the glass walls would fluoresce under certain conditions, and when a small amount of gas was left inside the tube, the gas molecules would glow along the path of the previously invisible cathode rays, as they came to be called. This was the precursor of the television picture tube, though most observers probably did not see that far ahead. The high voltage caused the loosely bound electrons to "boil" off the cathode (negative terminal) and thus be attracted across to the anode (positive terminal), and once these cathode rays were subjected to a magnetic field, they were observed to deflect from a linear path. The direction of this deflection indicated that the rays were negatively charged, and by measuring the radius of curvature involved in the deflection, Thomson was able to measure the charge to mass ratio of the cathode rays. Roentgen discovered that when targets are placed in the path of these rays, highly penetrating *uncharged* radiation was sometimes produced, radiation that could penetrate the body and picture the bones inside: X-rays!

We will look into this in more detail when we study the four fundamental forces, but at this point in the progress of physics we had found a way to experimentally determine the charge to mass ratio of the electron. It was left to Robert Millikan, in the first part of the 20th century, to design an experiment that would give us the charge on the electron, which then also yielded the (rest) mass of the electron using the ratio established by Thomson. Millikan was an extremely clever experimentalist, and his design to find the charge on an electron was conceptually straightforward and simple to understand. The implementation of it, however, was an entirely different matter. The setup called for tiny oil drops to be passed through a very thin tube – much like the atomizer of a perfume bottle. As the oil drops passed through the tube, the rubbing on the material of the tube caused the oil drops to become charged by friction in the same way as rubbing one's feet on a thick carpet. The charged droplets were then sprayed into a chamber that contained a parallel plate capacitor, a device with two metal plates separated by a small distance and connected to a battery or some DC voltage source. The top plate of the capacitor was charged positive by the battery, and the bottom plate negative. In this way, it was discovered that with the correct voltage, the oil drops could be suspended motionless between the plates. The equilibrium was established by balancing the downward force of gravity with the upward electrical force due to the charged plates acting on the oil drops. Of course, this implies that the oil drops must be negatively charged, since opposite charges attract and the top plate was positive (and the bottom plate negative – like charges repel). The electrical force was known to be equal to the charge on the oil drop times the electric field produced by the capacitor, and the electric field produced by a parallel capacitor was known to be the voltage on it divided by the distance between the plates. The gravitational force was naturally just the weight of the oil drops: their mass times the acceleration of gravity at the Earth's surface. In shorthand, the equilibrium formula that keeps the oil drops motionless looks like this: $mg = QV/d$, where m = mass of the oil drop, g = the acceleration of gravity, Q = charge, V = voltage, and d = distance between the plates. Consider all these variables, keeping in mind that we are trying to measure the excess charge on the oil drops gained from moving through the thin capillary-like tubing. Which of the variables would be the hardest to measure experimentally?

In practice, the oil drops were so tiny that a microscope was necessary to observe them in the apparatus. Years before he had tried using water, but it evaporated too quickly for enough accuracy to be attained. Millikan recorded data on thousands of these oil drops, peering through his microscope, a feat which did not do wonders for his eyesight. It turns out the hardest part of the observation was nailing down the mass of the oil drops, as they were incredibly small. The acceleration of gravity was well known from kinematics (free-fall experiments, pendulums, and much more), and the voltage and distance could be adjusted and easily measured. To determine the mass of the oil drops, since there were no remarkably precise balances at that time that could be used, Millikan developed an ingenious method based on air resistance, Stoke's Law, and the idea that with the voltage switched off the oil drops would just fall and reach terminal velocity quickly from the air resistance encountered during the fall. Using this method, and unfortunately a slightly erroneous value for the viscosity of air, Millikan could deduce the mass of the oil drops. The rest of the mathematics was then easy to do: plug in the measured variables and solve for Q, the excess charge on the oil drops.

The results were revolutionary and changed the course of history, because once the charge on an electron is known, so is its mass, and both are necessary components of developing a complete theory of the atom and all electronic devices. The experimental data revealed the following: every single oil drop observed either had a charge of around 1.60×10^{-19} Coulombs or some integral multiple of that elementary charge. No charge smaller than 1.60×10^{-19} Coulombs was ever observed, and no charge somewhere between, for example a charge of 2.0×10^{-19} Coulombs, was ever observed. Thus, Millikan had found the charge on an electron, the elementary charge often denoted as "e", and he had also discovered that charge, just like light energy emitted by blackbodies, is quantized: it comes in "lumps" of the elementary charge. This should make sense, since either the oil drop should have an excess of one electron, two electrons, ten electrons, but never two and a half electrons! The concept of fractional charge and quarks lay much further down the road historically, and when we investigate the nuclear force, we will return to that idea. This seminal work, along with contributions to the photoelectric effect, earned Millikan the Nobel Prize in physics in 1923. The importance of knowing the value of the elementary charge cannot be overstated – nearly all the progress in physics in the 20th century depended on it. But the story surrounding the physics of moving charge goes much deeper still, as it turns out its very movement produces some remarkable effects in the universe.

By the year 1800 AD, Volta and Galvani had invented the battery, and electrostatic phenomena had been well known for thousands of years. As we have seen, Coulomb and Franklin had done some work with electricity; it was also the case that Gilbert and others had done some work with certain materials that showed yet another force of nature, considered separate and distinct, called magnetism. Biot and Savart, two French physicists, developed a law similar to Coulomb's Law that was applied to two charges, except they coined a law that was for two magnets: like poles repel, unlike poles attract. *In all of this, it is important to remember magnetic poles are not electrically charged one way or the other, they are electrically neutral.*

In the early part of the 19th century, electricity and magnetism were thought of as entirely separate phenomena. For thousands of years people had known of magnetic materials (discovered in Magnesia, hence the term) as well as the electrical charging of straw and other materials by friction and so on. But these two processes were considered completely independent of one another until new technology allowed electric current to be studied, and then the world was never to be the same. Hans Christian Oersted, around 1821, discovered through the use of Volta's battery and some wiring that electric current going through wires causes magnetic compass needles, placed near the wires, to deflect. The conclusion he reached was inescapable: *electricity produces magnetism*. In the parlance of Faraday, electricity produces magnetic fields through space which then cause magnetic materials to respond. Exactly why this happens was not understood, but the experimental results were undeniable. There immediately followed two consequences of Oersted's discovery:

1. *A current-carrying wire acts like a magnet,* so if placed near a permanent magnet (like the kind that attach to refrigerators, for example) the wire should experience a force just like two magnets do. This is the force, transferred to an armature (coil of wire placed inside a permanent magnet or around another coil of wire), which creates a torque that makes electric motors run (saws, blenders, every single motor in existence), allows electric meters to register current and voltage, lifts cars in junkyards, removes metal from scrap heaps, opens apartment building doors, engages a starter motor in a car, causes woofers and tweeters to work in speakers, and thousands of other applications!

2. Two current carrying wires actually work like two magnets, and if oriented properly can cause a force of repulsion between the two, which can be used for magnetic levitation! This is the principle behind some of the MAGLEV trains in operation today!

Now if electricity can produce magnetism, it was reasonable to think that magnetism might be able to produce electricity. This looks simple, but it was hard to prove, and due to the slow pace of communication at the time, it wasn't until 1832, about a decade later from Oersted's discovery, that Michael Faraday uncloaked another secret of nature: a *changing* or *moving* magnetic field (in a particular way) can produce current in a wire! Stationary magnets or magnetic fields, like holding a bar magnet inside a coil of wire (coils of wire wrapped around a core are called solenoids), will not produce electricity. But if one moves the magnet through the coil, magically there is electric current in the wire! Thus, *changing* magnetic fields produce electricity, and this is known as Faraday's Law (of induction). The implications of this discovery, coming about ten years after Oersted (remember there was no radio, TV, telephone, and so on back then so news traveled slowly), were absolutely enormous: this is the principle behind all electric generators, telephones, telegraphs, and thousands of other inventions commonly used today. An electric generator is just a motor run backwards. Whereas a motor takes an electrical input and produces a mechanical output, a generator takes a mechanical input (turbine made to spin by pressurized steam or moving water) and generates electricity. This allows the distribution of electrical energy to everyone, and this is how everything electrical (other than batteries) in your home gets its energy!

As we have discussed, Faraday invented the concept of "field" to explain how two objects not in physical contact with one another could influence one another nonetheless (called "action at a distance"). Normally, in order to move something, we have to physically touch it to cause it to change course from a straight line or from rest. Remember there are three ways to accelerate an object: speed it up, slow it down, or change its direction. In all three cases, a net force is required to cause the acceleration. Forces exist only if there are two or more objects that interact with the field that each produces. Thus, the Sun's gravitational field interacts with the Earth's gravitational field and the force of gravity that ensues between the two causes the Earth to orbit the Sun. In the same way, positive charge produces an electric field, and if another electric field is present due to another charge (positive or negative), then there will be an electrical force between the two charges (likes repel, opposites attract). This extends to magnetic fields as well, whether produced by two permanent magnets or by two current carrying wires (which each produce a magnetic field). Now for another unsolved mystery in physics.

In electrical work, we learned that positive charge is pictured with electric field lines leaving the positive charge and with electric field lines entering negative charge. In magnetism, we picture the magnetic field lines as leaving the north pole of the magnet, and thus magnetic field lines enter the south pole of the magnet. (Remember this does not mean the poles of the magnet have a net electrical charge: in fact, the poles are electrically neutral!) We know that electrical charge can be isolated. That is, positive charge can exist without any associated negative charge having to be nearby along with it. Isolated protons, for example, occur throughout nature, as do isolated electrons. *The same is not true for magnets*! It turns out we have never found a magnet, or any magnetic field, whereby there is an isolated north or south pole only. If there is a north pole of a magnet, there must be a south pole nearby associated with it. Magnetic monopoles do not seem to exist, even though we are still looking for them and theory predicts they *should* exist, or at least no law of physics prohibits their existence. Either they must be exceedingly rare or exceptionally difficult to detect! The search goes on.

One might think that if you cut a bar magnet in half, you then will have a north pole isolated from a south pole. This is not what happens. When you cut it in half, each half has a north and a south pole, and each will be of the same strength provided you do not mess up the "domains" of the magnet by heating it or dropping it. Magnetic field lines always seem to go out of the north pole and end up in a south pole nearby – continuous loops if you will. Suppose we keep cutting the magnet until we end up with a single atom: do we now have a magnetic monopole? No! It turns out all atoms have a north pole and associated south pole – this is what allows us to use MRI scans medically: atoms are like tiny bar magnets! Well, what about the electrons within the atom? Nope! It turns out electrons have a north and a south pole, too. You can guess what the situation is for protons as well – they are tiny magnets, too, with a north and south pole. Nobody has ever found a particle or object having only one magnetic pole, although the search is on in earnest and has been for years.

Why are some materials magnetic (iron, cobalt, nickel) while others are not (sulfur, oxygen)? The answer to this question could not be arrived at until we understood the atom much better, which takes us well into the 20th century. First, there must be some moving charge or current to cause the magnetism, and in magnetic materials the electrons in the atom align their "spins" with an external magnetic field, causing the material to become a "permanent" magnet. This is actually a misnomer, since all magnets, like batteries, "run down" over time. While their domains may be "frozen" into a magnet for a time, this is not truly permanent. Heating or dropping a magnet can cause it to lose its magnetism, as can a number of other processes. We will return to electron spin at the end of the book, but for now one may consider it as analogous to a spinning sphere with two possible orientations, called spin up and spin down, but we should be very careful to note that an actual spinning electron (like a spinning top) is a figment of our imaginations, a tool used to try to visualize the inner mechanisms of the atom.

Some other developments in the 19th century were occurring simultaneously: Andre Ampere developed a mathematical formula that describes the strength of magnetic fields around current carrying wires, which came to be known as Ampere's Law. Adams and Leverrier discover Neptune; in fact, many of these planetary discoveries followed directly from predictions made by Newton's Universal Law of Gravitation. Returning our focus to this "new" phenomenon of magnetism, a central question and debate raged on: what causes the Earth's magnetic field? We know there is one, otherwise compass needles would not work! We are not exactly sure, but read on for some interesting ideas and things we do know about it.

The origin of the Earth's magnetic field is believed to be moving charges, in other words convection currents, in the Earth's outer core. From seismic studies and the "shadow zone" created by Earthquake waves traveling through the Earth's inner core, we can deduce densities of the materials at or near the core, and from that conclude that the inner core is composed of iron, nickel, cobalt – all elements with an electronic structure such that they are readily "magnetized" if there is a current in the region. It is these convection currents, highly variable on a geologic time scale, that appear to cause the magnetic field of the Earth. We know from studying geological formations, rocks, and fossilized samples that the Earth's magnetic field has reversed itself many times in our history, and has therefore passed through "zero" points when there was little or no magnetic field produced by the planet. As you will see, this has tremendous implications for us biologically and in terms of species survival. Apparently, these convection currents can halt altogether at some points in our history, and reverse direction as well.

At the moment, the strength of the Earth's magnetic field is about a micro Tesla (Teslas are the metric SI unit for magnetic field strengths). While this may not seem like much, it is enough to navigate by and it will register on even the least sensitive compass needles. The strongest magnetic fields we can make (in MRIs and at the CERN particle collider) can reach values of 10 Teslas or more. The Earth's magnetic field is not aligned with our geographic poles; a line drawn from magnetic north to south does not align itself with the rotational axis of the Earth. At the moment, the angle of declination, which is the angle between the magnetic pole axis and the geographic axis, lies somewhere between 10 to 20 degrees. This declination must be taken into account when navigating long distances, as magnetic north is not true north. The dip angle is about 70 degrees below the horizontal, and this occurs because the origin of the Earth's magnetic field is in fact far beneath the surface of the Earth in the outer core. Compass needles point downward for that reason. But what happens when there is no magnetic field?

High energy, charged radiation called cosmic rays are constantly streaming from the Sun towards the Earth. This moving charge produces its own magnetic field (Oersted!), and interacts with the Earth's magnetic field to produce a magnetic force on the cosmic rays. This force is maximized when the particles move at right angles across the Earth's magnetic field lines, and slowly drops to zero at the point where the charged particles move parallel with the Earth's magnetic field lines. There is a visual we use in physics, called the "right hand rule" or RHR, which applies to the movement of *positive* charges or the direction of electric current. (For negative charge movement, the simplest way is to apply the same rule to one's *left* hand.) Although there are variations of it, I have found the most useful application of the RHR to be as follows: stretch your right hand out so that the thumb is perpendicular to the outstretched fingers and your palm is facing outward – like you were raising your hand but with the thumb pointed parallel to the ground. Your thumb represents the direction the positive charge is moving in, your fingers represent the direction of the magnetic field lines present in the region, and your palm represents the magnetic force that pushes the particles away from their original linear path into a circular path, changing their direction of travel but not their speed. In general, the direction of the charged particle's velocity is perpendicular to the magnetic field lines and the magnetic force thus produced is perpendicular to both of those! Hence, the charged particles (cosmic rays) are deflected around the Earth in belts called the Van Allen Radiation Belts at locations where they cross over the Earth's magnetic field lines, and these locations are more toward the equator than the poles. This deflection does not occur as much at or near the geographic poles, since here the particles are traveling nearly parallel to the Earth's field lines, and so the magnetic force drops to zero. Hence, if conditions are right, we can observe the "northern lights" as charged particles dip down into the atmosphere and put on a light show. One can also see, then, that if there is zero magnetic field for the Earth, cosmic ray bombardment is greatly increased at the surface of the planet, and this may be the cause for periodic mutations and extinctions throughout our history.

While all of these discoveries are going on with electricity and magnetism (Oersted, Faraday, et al), there are also major breakthroughs in thermodynamics (Carnot, Joule, Maxwell, Boltzmann) and Mendeleev is developing the periodic table of the elements. While the existence of atoms is a hot topic of debate in some circles at this time in the 19th century, there is also debate over the nature of light (wave versus particle – recall Young's double slit versus Newton's particle model), and there are great mathematicians like Hamilton and LaGrange who are developing new and powerful ways to do Newtonian Mechanics. The Industrial Age and western expansion are in full swing; classical and romantic eras in art and music flourish, and as you will see, the world of physics becomes largely a coronation of the work done by Newton hundreds of years prior to these discoveries, as well as the work that will be done by James Clerk Maxwell, largely during the Civil War era of the 19th century. Maxwell is easily the greatest physicist of the 19th century, at least according to most historians of science, with his work in thermodynamics but especially with his revolutionary synthesis called electromagnetism.

There were many other developments in electricity and magnetism that followed from Oersted and Faraday: Lenz's Law of self-inductance is but one example. You may have noticed that sometimes lights will dim momentarily when a large motor turns on, like a furnace or refrigerator. Then, once the motor is up to speed, the lights come back on at full brightness. (In modern wiring, this should not happen since lights should be on a separate circuit from large motors!) What causes this momentary dimming? Follow the chain of logic below, keeping in mind all of it flows from the discoveries of Oersted and Faraday:

1. First, electric current is supplied to the motor's armature, which is a coil of wires wrapped around a cylinder that can spin inside the motor.

2. The current in the armature produces a magnetic field, and there is another magnetic field built into the motor either through a permanent magnet mounted in the motor or through field coils that act like a magnet when there is current moving through them. These two magnetic fields interact to produce a torque which then spins the armature. The initial current required to get the armature spinning is quite large, hence the lights dim if they are on the same circuit.

3. The spinning armature now acts like a generator! Since it is a moving or changing magnetic field in itself, it produces a back voltage, called back EMF, that acts to oppose the original current coming into the motor. This backwash, if you will, acts to limit the current, and once the motor's armature is spinning at full speed, the generated back EMF limits the incoming current to the point where the lights come back on to full brightness. This back EMF is Faraday's Law in action, and is really a consequence of the Conservation of Energy that we will investigate in the final section of the book, but here is a brief synopsis of the ideas involved.

To see why this is based on the Conservation of Energy, imagine the current generated by the spinning armature *added* to the current coming into the motor. One can swiftly see that there would be an ever increasing current and an endless supply of energy so we could eventually unplug the motor and have it run by itself for free! Clearly this doesn't happen, and that is because it can't happen, because we can't get energy "for free", one of the most important ideas in all of physics.

The state of electricity and magnetism in physics around 1850 was as follows. We knew that stationary electric charge produces an electric field, and this could be described by Coulomb's Law. We knew that moving charge or current produces a magnetic field; we knew that changing or moving magnetic fields, if oriented properly and moved through coils of wire, produce electric current or moving charges. We knew that stationary magnets produce only magnetic fields. We knew *how* to generate electricity, but we had no real idea as to *why* this all works, since we had no understanding of things at the atomic level, and in fact the very existence of atoms was still in much doubt. Though we could generate electricity, we had no way of transmitting this electrical energy over long distances without losing it all to heat energy within a matter of miles due to the resistance in the wires. At this juncture, Maxwell enters the picture in a way that changes human history forever, and the modern era of communication, transportation, and so forth soon follows.

Maxwell, in an extraordinary example of "seeing the big picture", summarizes all of electromagnetic theory into what has become known as Maxwell's Equations. You will see those on coffee cups and tee shirts, a bunch of mathematical symbols culminating in the words "and then there was light". Here they are in words:

1. A generalized form of Coulomb's Law called Gauss' Law, the definition of how electric charge produces an electric field, and hence how electric charges interact to produce electric force.

2. Biot Savart Law, similar to the Coulomb's Law but for magnets: the definition of how magnets produce magnetic fields, and the fact that magnetic field lines always go in continuous loops from north to south pole.
3. Faraday's Law: changing or moving magnetic fields produce electric fields which in turn produce moving charge or current electricity.
4. Oersted's Law: moving electric charge or current produces magnetic fields (and Ampere quantitatively analyzes the strength of these fields).

Now if this is all Maxwell did, you might reasonably wonder why he is often rated as the greatest physicist of the century. Here is the reason: to these four equations, Maxwell added a crucial and revolutionary element: *perhaps changing electric fields produce changing magnetic fields, which in turn produce changing electric fields, which produce changing magnetic fields*…you may see the pattern here...and the result is an electromagnetic (EM) wave that spreads outward in spherical wavefronts through space, like ripples in a pond when a large rock is dropped into it. Maxwell then develops the mathematics to calculate the speed of these waves, and the result looks like this:

$V = 1/(\varepsilon_0 \mu_0)^{1/2}$ where ε_0 comes from the electric field predicted in Coulomb's Law and equals 8.85×10^{-12} in metric units and μ_0 comes from the magnetic Biot Savart Law and $= 4\pi \times 10^{-7}$ in metric units. V = the speed of the EM wave, and the power of ½ means take the square root.

You should pop these numbers into a calculator and do the math, then stare at the result and maybe you might experience a fraction of the excitement that must have filled Maxwell when he first crunched the numbers. Plugging in the values for the two constants, and following the formula for finding the speed V above, Maxwell found the result: $V = 3 \times 10^8$ m/s!!!!

He was enough of a physicist to know that this number is hugely significant in physics: it is of course, the speed of light. He concluded that *EM waves are, in fact, light itself!* Thus, a rubber rod charged electrically, a bar magnet, and a beam of light were all part of the same thing: electromagnetism. This is a gigantic synthesis of physical phenomena, and its importance cannot be overstated. The ramifications of this discovery lead directly to radio, TV, computers, and much more. The world shrinks dramatically once we can communicate almost instantaneously.

Recall at this time Young had already done his double slit experiment showing that light exhibited interference and thus had wave properties. Several things ensue from this outcome: if light is a wave, where and what is the medium that transports this wave? We have seen that this supposed "ether" that transmits light waves has never been detected, and the null result from the Michelson and Morley Experiment is one of the most famous "finds" in all of experimental physics. Eventually, Einstein uses Maxwell's ideas and develops two postulates, one based on Galileo's work and the other on Maxwell's work, which lead directly to the Special Theory of Relativity and the idea of space and time being relative and the speed of light being absolute (rather than the other way around, which is what Newton and others thought). We have also seen, through the photoelectric effect and some other ideas brought to the fore in the 20th century, that light has very definite particle properties as well, and so has a dual nature, and from there it was a short leap to suppose that *all things* have a dual nature in the universe, and which one is observed (wave or particle) depends on which experiment is done, and also on the very act of observation itself! The universe we live in is as strange as it is miraculous.

Some practical ramifications now follow from Maxwell's work, occurring shortly after the American Civil War:

1. If accelerating charge produces EM waves, and the electric and magnetic fields are perpendicular to each other and the direction of the wave travel, then light is a transverse wave and can be polarized (cancel out the electric field and/or magnetic field with a fine mesh that looks like a tennis racket under a microscope...fields cannot travel through the mesh if oriented properly...light is blocked. Think: sunglasses!)

2. So...if we rub a comb through our hair and charge it electrostatically, then move it up and down vigorously so as to accelerate this charge, we generate an EM or light wave? Yes! But the frequencies we do this at are so low that we cannot see this wave as visible light. Visible light has frequencies in the range of 10^{14} cycles/sec, so we would need a lot of coffee to move the comb up and down that fast! However, in the 1880s, Heinrich Hertz experiments with Maxwell's theory and builds an antenna that can accelerate electrons up and down the antenna with very high frequencies, and thus he succeeds in sending out radio waves (light waves of a specific frequency in the megahertz range typically) and also visible light waves. This is the invention of the radio, though Tesla and Marconi lay claim to the patent as well. Unfortunately, Maxwell did not live to see this vindication of his theory. Marconi and others go on to perfect the radio: a receiving antenna detects that EM wave being sent, and through resonance and electric circuits, the wave of a particular frequency is selected out and amplified. This electric current then travels to a

speaker with magnets, which vibrate in exactly the same way the microphone's magnets did to produce the sound that was generated! All of EM theory is in play here. Sound energy converted to EM energy…sent through space as EM waves…received by Faraday's Law at the receiving antenna…converted back into sound by speakers via Oersted…incredible stuff, but not magic.

3. Tesla and others perfect/develop AM/FM tuning, radio, and much more as time progresses. The telegraph and telephone are in the making, along with every other kind of EM device you can think of…life is never the same.

Let's return to a problem we cited earlier that is very practical indeed: how do we transmit electrical energy over long distances so that people can have electricity in their homes, offices, and so forth. This is how that particular problem was solved, and not without considerable acrimony between the two major players: Edison and Westinghouse. We generate electricity through Faraday's Law: a turbine (huge coils of wire like a motor's armature) is made to spin through some mechanical means: steam from burning wood, coal, oil; steam from nuclear fission; water or wind or tides hitting the blades of the turbine. These spinning coils passing through a magnetic field then have current/electricity induced in them (one can either move a magnet through the coils or the coils through the magnet…it's easier to do the latter), and the electrical power is then transmitted to those who need to use it. The problem is, huge currents produce a tremendous amount of heat in the wires that transmit the electricity, and most of the energy ends up as heat rather than electricity in a matter of a few miles of wire. How is this engineering problem solved?

Many ideas were put forward, perhaps the most bizarre of which was Tesla's idea to generate such huge voltages in a region that the electrical energy alone would be enough to run any electrical device within a few miles of this giant Tesla coil without any need for wires at all. This wasn't very practical, and in fact got Tesla in a lot of trouble in some locales! Here are the facts leading up to the solution employed, and the one we still use today, 150 years later:

1. One way to reduce the heat loss in the transmission process is to reduce the electrical resistance of the wires. Recall that the resistance of the wires is related to the material, the length, and the cross-sectional area as well as the temperature. The length can't be altered since there is a set distance from the turbine generating the electricity to the city that needs the electricity. The cross- sectional area can be increased, but there are limits to how much this can be done since the size of the wire can become too big to support realistically. The temperature outside cannot be controlled easily or at all in most cases; the type of material can be controlled so we need to use a good conductor. Silver and gold are excellent conductors, but hundreds of miles of either would be a problem since people would cut them down and sell them for profit! Copper is the next best alternative, but we now use aluminum because it is cheaper (more plentiful, too). The loosely bound electrons in metal make them ideal for transporting electrical energy. So, now we have aluminum power lines and this helps with the problem, but does not solve it! Currents are still too high.

2. Power loss in a circuit depends on the resistance of the line and the amount of current *squared*, so the amount of current is tremendously

important since it is squared. How do we then keep the current to a minimum and thus minimize heat/power loss? Recall that electric power is equal to Voltage x Current, so if we can keep the voltage very high, then the current can be kept relatively low. So, we need to enable the power from the turbine to be transmitted at very high voltages and low current over the long distances required, then at the receiving end we need to *transform* the voltages back down and the current levels back up so that the electricity can be used. Otherwise the huge voltages at the receiving end will blow things up, like lights, TVs, …you get the picture.

3. Thus, we need an electrical *transformer* that can transform voltages and currents accordingly. A step-up transformer (or series of them) is required at the generating station to get the extremely high voltages and low currents needed for efficient transmission, then a series of step-down transformers is needed at the receiving end (homes and offices) to get the 240 volts and higher currents needed for the operation of household and other items.

4. It turns out the transformer only works on AC electricity (alternating current), and this is why/how Westinghouse won the battle over Edison: Edison was a DC (direct, steady current) advocate, Westinghouse advocated the use of AC. By changing connections at the turbine, one can generate either kind of electricity, but to transmit the electricity with high efficiency and low power loss, we need transformers, and they only work on AC electricity. To understand why only AC, we dig a little deeper.

The following sequence of ideas outlines the operation of electrical transformers. Since this is the device that made it all possible, it is important to understand the physics of its operation. Three key facts emerge:

1. The voltage in a portion of the transformer is directly proportional to the number of turns of wire wrapped around it. The more wire turns, the greater the voltage.

2. Only *changing* magnetic fields produce electricity, so *changing or alternating* current is required to produce those changing magnetic fields, hence only AC will work!

3. Power IN must equal Power OUT (assuming very little power loss)

In a typical transformer, pared down to its essentials for simplicity, this is how the physics works. We will use a step-up transformer for our example, but the same principles apply for a step-down transformer (just reverse the sequence).

1. AC electricity is generated at the turbine and supplied to the primary or input side of the transformer. This AC produces a changing magnetic field within the transformer, and this changing field travels through the secondary coil. This is essentially Oersted's discovery in action: electricity produces magnetism, in this case, changing current and thus changing magnetic fields.

2. The changing magnetic field traveling through the transformer's iron core moves through the secondary coil on the output side, and this produces electricity due to Faraday's Law. Since we have

increased the number of turns of wire on the secondary coil, the voltage will be proportionally higher than that of the input side.

3. Since Power = Voltage x Current, and we have increased the voltage dramatically on the output or secondary side, the current on the secondary side must decrease (power in must equal power out, or close to it – transformers do suffer a little power loss).

4. Now that we have made the current relatively small, we can transmit power over long distances with small heat/power loss in the transmission lines. Typical efficiencies run over 90%!

5. At the receiving end, we merely reverse the process with a series of step-down transformers (far fewer coils on the output or secondary side of the transformer) until we reach 240 volts, higher currents, and thus deliver the energy to homes and offices.

Transformers "hum" due to the 60 cycle/second (Hertz, abbreviated Hz) and other factors; they are cooled down internally through a variety of means; this used to be with PCBs but the environmental impact of those chemicals has made such means of cooling far less common and in fact, prohibited in most cases. Since electricity in the United States is generated at 60 Hertz, the turbines spin at 3600 RPMs. The generation is done by AC, so 120 times a second there is zero current in the wires, but this alternating cycle is so fast that it cannot be perceived and has no real downside. Thus, the upshot of all of this is that the world shrinks in remarkable ways: telephone, telegraph, computers, internet, radio, TV…all the appliances we use and all the electricity we use can be traced back to these developments. Communication in many forms becomes instantaneous, and in such cases wires are not necessary to do so!

Let's take a final look at EM phenomena by summarizing some aspects of it and by looking at the entire EM spectrum possible: all the wavelengths of light we know and what they are in practical terms.

1. All EM waves travel at the speed of light, because they *are* light: roughly 300,000,000 meters per second in a vacuum.
2. There is no medium we have ever found that transports these EM waves.
3. These waves are transverse in nature, in that the electric and magnetic field vibrations which propagate the waves run perpendicular to the wave's direction of travel. All three, the electric field, the magnetic field, and the velocity of the wave, are mutually perpendicular vectors.
4. The speed of EM waves, in other words light itself, does not depend on the reference frame of the observer. It is absolute and the same to all observers.

The EM spectrum is charted on the following page.

TYPE OF EM WAVE	frequency in Hz	wavelength in metric	speed (c)
RADIO WAVES	$< 3 \times 10^{11}$	Greater than 1 mm	300,000,000 m/s
MICROWAVES	$3 \times 10^{11} - 10^{13}$	1 mm to 25 μm	300,000,000 m/s
INFRARED	$10^{13} - 10^{14}$	25 μm – 2.5 μm	300,000,000 m/s
VISIBLE	$4-7.5 \times 10^{14}$	750 – 400 nm	300,000,000 m/s
UV (a, b, c)	$10^{15} - 10^{17}$	400 nm – 1 nm	300,000,000 m/s
X RAYS	$10^{17} - 10^{20}$	1 nm – 1 pm	300,000,000 m/s
GAMMA RAYS	$10^{20} - 10^{24}$	$< 10^{-12}$ m	300,000,000 m/s

Some facts:

1. At the UV stage and at higher frequencies, EM Waves have the ability to penetrate materials, including humans. The higher the frequency of the EM radiation, the greater the energy: recall $E = hf$ from Planck's blackbody radiation.
2. Note that the visible portion of the EM spectrum is incredibly tiny…a small slice of the universe is visible to

us, perhaps that portion which was necessary for our survival over millions of years.

3. Note that *all* of these are considered to be light. The higher the frequency, the shorter the wavelength, and the shorter the wavelength, the more "particle-like" they become.

As one final example to drive some of these points home, let us engineer the following process. We run electrons up and down a wire (antenna) a million times a second or more, generating an EM wave, and claim credit for the invention of the radio! The credit for this invention is still an ongoing dispute among Tesla, Hertz, Marconi and others. Then we can speak into a microphone that has a tiny magnet on the inside, and the vibrations of the sound waves we create causes the magnet to move in synch with the sound wave, and moving magnets produce electricity (Faraday). This electrical signal is then mixed piggy-back style onto the radio wave we just discovered, either by amplitude modulation (AM radio) or frequency modulation (FM radio). The whole package is sent through space at the speed of light, since radio waves are light waves, and hits the antenna or receiving platform at your house. You now use inductors and capacitors to "tune" the circuit so that it will only resonate at the frequency of the station you want to hear, and so just that signal is selected out among the many radio waves always in the air. Now, you get rid of the radio wave that transported the electrical signal that carries information of the sound wave you created at the microphone. The electrical signal causes magnets in the woofer and tweeter of the speaker system to vibrate in exactly the same way as the microphone magnet did, reproducing the sound wave that created it in the first place. In an instant, music, words, images, information, and all the rest is transported around the world at the speed of light. Magical indeed, but not magic: it's physics – the laws of nature uncovered and in action. Television, computers, iPhones, all of it...comes back to EM theory and the discoveries made in the 19th century.

We have arrived at a significant milestone in our journey, though it by no means signifies the end of all the "vistas" one might encounter along the way to analyzing the four basic concepts in physics: space, time, mass, and charge. It is worthwhile, however, to pause for a long moment and consider the views we have afforded ourselves in the process. Though quite fundamental to all of physics and in much of our experience as human beings, as we dig deeper into the meaning of these four concepts, we enter fantastic and undiscovered terrain all along the various paths. One of the challenges in working through this material is deciding what paths to take, and indeed how far along each of the "sidebars" one wants to go. Any one of these sidetracks could fill volumes; many require a lifetime of study to master, and even then, it is an evolving enterprise.

It is said that true enlightenment is not the arriving at the summit of a mountain, it is in the absolute awareness of the beauty and the work in each step of the hike. In any major undertaking, whether it be writing a book, winning a championship, designing a bridge, running a bank, painting a masterpiece...whatever enterprise requiring a steadfastness and sense of purpose, one would do well to remember to take joy along the paths taken, together with the inevitable pain that also exists. I hope this section regarding the four basic concepts in physics has raised more questions than it has answered, for that is one of the main goals of this book. I hope it has raised your sense of wonder regarding the universe we live in. We move on now to explore the four fundamental forces that govern all interactions in our universe.

The Four Forces: Gravity, Electromagnetism, Weak Nuclear, Strong Nuclear

The above four forces, or interactions as they are more often called these days, account for every single phenomenon observed in our universe to date, at least that is what the data seems to indicate at this writing. We understand some of these forces better than others, but we have uncovered many new vistas in our exploration of all of them. One of the main goals in physics has been to unify these four fundamental forces into one force – an interaction that can encompass all known observations. There has been some success along this path, but the ultimate goal of "grand unification" has yet to be achieved. It is uncertain if it ever will be, but the journey continues. We begin this section of the book by first defining what we mean by the term "force": how it is defined, measured, and operationalized. Then we will start our trek into the four fundamental forces with a deep look into the interaction we have known and studied for the longest period of time, yet paradoxically still remains as the most enigmatic in many ways: the force of gravity.

FORCE

Recall that in the field of physics called kinematics, the exploration centers on *how* things move: how far, how fast, how much time, and how the motion is changing if at all. As we have seen, space and time are the two basic concepts in physics that encompass all of kinematics. It is an extremely relevant field of study, in that the applications are rife: timing the yellow light at traffic signals, designing the length of runways safely, investigating motor vehicle accidents, launching spacecraft, designing rapid transit systems, trying to outrun a grizzly bear, and thousands of other practicalities. The variables in kinematics are given specific definitions and labels. Displacement, a vector, is the distance an object is from its original position together with its direction (angle) in relation to the starting point. Distance is a scalar quantity and just specifies how far an object has gone with no regard for direction. Both are typically denoted by dimensions of x, y, and z for the three dimensions we live in, although there are other coordinate systems which can be used to define points in space. The standard metric unit for both is the meter. Velocity is the speed of an object together with the direction of its travel, and thus is a vector. Speed is how fast something is going without regard for direction, hence a scalar. Both are typically denoted by the symbol "v", and the units are space/time, which in the metric system equals meters/second, and of course time is a scalar given the symbol "t". Acceleration is a vector quantity because it has direction, and it is important to note that *its direction is not necessarily in the direction of motion of the object.* For example, when an object slows down, the direction of the acceleration is opposite that of its direction of motion. It is also important to note that *acceleration is not how fast an object is moving, rather it is a measure of how fast an object is changing how fast it is going (or its direction or both)!* In other words, acceleration measures a change in velocity, its symbol is given as "a", and the metric units are meters/second/second, or as it is often written m/s^2. Keep in mind there are three ways an object can accelerate: speed up (hit the gas), slow down (hit the brakes), change direction (turn the steering wheel). In all three cases, we know we are accelerating because we can *feel* it! There must be a cause for this acceleration, and that is where we are heading now.

The field of dynamics investigates *why* objects move, or more precisely why objects may change their state of motion and undergo acceleration. In order to explore this terrain, the third fundamental concept, mass, becomes a critical component of the analysis. Galileo was the first to experimentally verify that there is a property of matter called inertia: the resistance to *changing* a state of motion. It is worth repeating that this is vastly different than Aristotle's contention that objects resist motion altogether, or stated another way, he posited that the natural state of objects is to be at rest. Galileo showed this idea to be incorrect by digging deeper into the patterns of motion seen in nature, and astutely observed that absent friction, the natural state of objects would be to continue doing what they are doing forever. There is still no verifiable *explanation* for this law of inertia, but it remains a cornerstone in the foundation of physics nonetheless. Changing a state of motion is another way of saying that an object is accelerating, and for that there must be a cause, and the cause involves the physics concept called force. When diving into the field of physics involving forces, all roads lead to Isaac Newton.

Newton was born on Christmas Day (depending on which calendar one uses) in the same year Galileo died, 1642. During his lifetime he was knighted, a member of Parliament, a Lucasian Professor of Mathematics, Master of the Mint (employed to catch counterfeiters, which he was quite successful at doing), President of the Royal Society, and of course he was counted by many as the greatest scientist who ever lived. Even to this day, the honor of being considered the GOAT for scientists is a tossup between Newton and Einstein. His father died just months before he was born, and eventually he was cared for by his maternal grandmother, as his mother remarried. When his stepfather died and his mother became a widow once again, it was the intervention of an educator that finally persuaded her to have Newton go back to school and not become a farmer. During the bubonic plague, college students were sent home to escape the ravages of the disease, and it was in this time period that Newton's intellectual pursuits flourished in rapid fashion. He developed the first successful reflecting telescope, studied optics (recall his particle or corpuscular model of light) and wrote papers explaining how white light could be broken into colors with a prism. Newton proposed an ether that permeated all of space (sound familiar?) which could transmit interactions through a vacuum across vast distances, a notion that would serve him well in the future. He was a dedicated alchemist his entire life, and most historians regard his religious beliefs as complex and not fully understood, as Newton guarded his private life carefully. He was involved in several fairly vicious disputes with fellow scientists and mathematicians: Leibnitz over the invention of calculus and Hooke over a variety of matters. Both feuds lasted until his adversaries died. Edmund Halley, of Halley's Comet fame, befriended him and helped him publish what is probably the greatest scientific work ever written:

Principia Mathematica. Newton remained a bachelor all his life, and gained much fame and many honors in his comparatively long life. He died in March of 1727 and was buried in Westminster Abbey, an indication of how revered he was across all of England.

I have left many of the details of his life out of this book, because there are dozens of excellent books which cover all the phases of Newton's life, and each one fills many pages with accounts of his extraordinary life. I invite the reader to investigate his life further, because in so doing, one may come to realize how every person, whether a genius or not, is inextricably connected to the lives that came before him or her, as well as heavily influenced by contemporaries. For this writing, we will focus on Newton's concept of force and how it manifested itself into his three laws of motion, and eventually the culminating glory of his work in my view: the universal law of gravitation. These laws of motion became the bedrock of what is now known as Classical or Newtonian Mechanics, and although one of these laws has been found to be incorrect at speeds approaching the speed of light, at temperatures approaching absolute zero, and inside the atom, much of his work is still used today in every field of engineering. The three previous exceptions cited were obviously not available to be tested in terms of experimentation, using the technology at the time in which he lived, and in most ordinary applications (designing roads, bridges, buildings, spaceflight, and so forth) Newtonian Mechanics is still employed with astonishing success. Relativity and quantum physics are needed to expand the circle of knowledge, encompassing Newton's work in the process as valid for "ordinary" speeds, temperatures, and masses.

The initial steps we will now take bring us to the field of mechanics and the concept of force as the primary means of changing the motion of objects. In the process, we will look at this concept of "action at a distance". How is it possible for one mass to be influenced by another mass when they are separated in space, sometimes by millions of miles? What is the invisible "rope" that spins the Earth around the Sun at breakneck speeds, holds the Moon in its orbit, and makes the apple fall to the Earth?

The essence of Newton's First Law of Motion had already been established by Galileo: his Law of Inertia. It can be written simply as follows: *objects will maintain their velocity unless an outside (external) force acts to change that velocity*. One begins to comprehend how vital it is in physics, and in all of science, to establish precise standards in our use of vocabulary, as well as the demonstrated necessity for precision in our standard units of measure. As we have seen, there hasn't been any satisfactory explanation for this property of matter called inertia, neither has there been any experimental violation of it. For the moment, it is a rule in nature's game of four (maybe more) dimensional chess, and it is applied uniformly with no apparent cause. Newton went much further, of course, and pinned down an operational definition of what is meant by an outside or external force.

His second law, the one that is used throughout thousands of applications in science and engineering unless we enter the three realms alluded to previously, brings in the third fundamental concept we analyzed: mass. Newton's reasoning was straightforward, and consisted of three logical lines of approach. First, forces or interactions often occur in large numbers simultaneously, so it must be the *net force*, or the sum of all the forces acting *on* an object, that changes the velocity of an object, and this occurs only if that net force does not equal zero. If all the forces acting on an object are balanced, then we can safely conclude that the object cannot change its linear or translational (how fast and in what direction it moves through space from one point to another) velocity. Recognize this could also mean it is at rest in a reference frame, which simply means the object must stay at rest if the net force is zero. This should make sense to the careful observer: if all the forces acting on a car add up to zero, it will keep doing what it is doing in terms of velocity! The second aspect of this approach posits that the greater the net force acting on an object, the greater the acceleration will be. Again, this seems logical to most observers: if there is a large imbalance in forces on an object, its velocity will undergo a large change. The third and final aspect involves mass: the greater the mass of an object, the less its acceleration will be, given the same net force acting on it. There is no mystery here: all other factors being equal, it is much harder to push a car forward than it is to push a toy wagon. More mass means more inertia, which means more resistance to changing the state of motion, which in turn implies more net force is required to do so. Synthesizing these three ideas leads to *Newton's Second Law: net force = mass x acceleration*, or in the shorthand of a formula, F_{net} = *ma*. Note that since acceleration is a vector, the net force must be also, and that *the acceleration therefore must be in the direction of the net force, and neither has to be in the direction the object is actually moving.* The units will be mass x space/time^2, or kilograms times meters per second squared. Physicists do not like writing all of that out every time, so we shorten the metric unit of force by calling it a Newton, in his honor of course and to save us a lot of time and ink! That's the up side to naming a unit; the down side is that people often forget all of these concepts in mechanics reduce to space, time, and mass, and one of the main points of this book is to remind you of that fact.

Newton's Third Law is the one most people are familiar with, but it is also the one that often leads to a lot of confusion. It is expressed most precisely by the following: *for every action force, there is an equal and opposite reaction force.* At first glance this law may seem obvious, but it is actually a deep insight into nature that has strong ties to one of the conservation laws discussed in the last section of this book. As one example, if I push on a wall, the wall pushes back on me with exactly the same amount of force. It *has* to be the same amount of force since it is one (and the same) interaction between me and the wall. If you are not convinced that a wall can push on you, try doing it with roller skates or on a skateboard. With little friction to impede your motion, you will fly backward away from the wall. Many people read the third law of motion and then wonder how anything ever moves, thinking the action and reaction forces must always cancel since they are equal and opposite. The critical observation here is that *action-reaction force pairs never cancel because they act on different objects. For two forces to cancel, the must indeed be equal and opposite but they must also act on the same object.* Therefore, whether or not I accelerate away from a wall when I push on it depends on the sum of all the forces acting *on me*, not the force acting *by* me pushing on the wall. Every interaction in nature occurs in such pairs; it is impossible to hit a wall hard with your fist without the wall hitting back just as hard, which is why it hurts!

When a bug smashes into the windshield of a moving car, it generally does not fare well as a result. Most people assume, logically to be sure, that the bug experiences a greater force than the windshield, and that is why it gets crushed. The third law teaches us that this is false: whatever force the bug exerts on the windshield is equal and opposite to the force the windshield exerts on the bug. Again, this has to be true since it is one interaction. The reason the bug gets clobbered has to do with mass, not force. When we invoke the second law, F_{net} = ma, the picture is much clearer: the bug has so little mass that its acceleration is huge, and *that* is what kills the bug. The windshield has a much greater mass, and therefore experiences only a tiny acceleration. One might visualize this by writing: mA (bug) = Ma (windshield). It is rapid changes in velocity, in other words large accelerations, that can be lethal. This is why we have padded dashboards, air bags, and safety nets when walking a tightrope: in each case, the time of impact is extended over several seconds, which causes the rate of change of velocity to decrease dramatically, meaning the acceleration is kept to small and hopefully safe levels. For the same reason, professional boxers wear padded gloves, as the use of bare fists at that level of competition would nearly always be fatal. On the flip side, if a karate expert is demonstrating how to break through a block of concrete, it will be observed that the delivered blow is sharp and done in an extremely short time period. This creates a lot of acceleration, and therefore a lot of force.

We will conclude this section on force with three examples: a horse pulling a wagon, a tug-of-war, and propulsion in spaceflight. The forces acting *on* a wagon being pulled by a horse, assuming level ground, include the tension in the reins caused by the horse pulling forward on the wagon, the force of friction the ground exerts on the wheels of the wagon, the pull of the Earth on the wagon, and the support force (called the "normal" or perpendicular force) of the ground pushing up on the wagon. Assuming the vertical forces are balanced and cancel (otherwise the wagon falls through the ground or rises up from it), we are left with the conclusion that the wagon accelerates forward if the tension in the reins is greater than the friction on the wheels. We can apply the same analysis to the forces acting *on* the horse. There is the tension in the reins caused by the wagon pulling back on the horse. This is the reaction force to the horse pulling on the wagon, and make note of the fact that it is the very same tension since this is one interaction through one set of reins. The force of the ground pushing forward on the horse as the horse gains traction by pushing backward on the ground – this is how we walk! The vertical forces of the Earth pulling down on the horse and the support of the ground pushing up on the horse complete the "free body force diagram" showing all the forces acting on the horse. Therefore, the horse accelerates forward if the traction it gains (the ground pushing forward on his hooves) exceeds the tension in the reins pulling back on the horse. *All mechanical systems can be analyzed in this manner: draw a free body force diagram of all the forces acting, then apply Newton's Laws of Motion to determine what effect all the forces will have on the motion of the object.* The analysis is not complete unless we also consider the effects of rotation, which we will deal with shortly.

In a tug-of-war, the two ingredients you want to have to win the battle are mass and traction. Arm strength will not matter, since the tension in the rope pulling each team has to be the same! More mass and better traction will lead to less acceleration, which of course means victory. When analyzing spaceflight, the key point is that expelling gas out of a rocket means the gas must react by pushing the rocket forward. Robert Goddard, a pioneer of rocketry in the 20th century, was widely ridiculed for suggesting rockets could be propelled in outer space. The argument was that since there was no atmosphere to push off of, rockets would not work once outside the thin envelope of air surrounding the Earth. Goddard thought differently, and so did the laws of physics! Action-reaction is not confined to the Earth: it works everywhere. In fact, rockets fly more efficiently in outer space because there is no air resistance holding them back. Once the rocket pushes gasses at very high velocity out the back, the reaction must follow that the gasses push the rocket forward. A gun firing a bullet will recoil no matter where it is located!

Newton's Three Laws of Motion apply to a vast array of phenomena, and they set the table nicely for our exploration into the force of gravity. The last step to complete before doing so is to look at how systems rotate in space, rather than just translate or move from one point to another. In general, objects have six degrees of freedom. They can translate along the x-axis, the y-axis, or the z-axis. They also could rotate in the xy plane, the xz plane, or the yz plane. As a visual, hold a ruler in front of you and consider the ways it can be moved through space. We could move it back and forth horizontally (x axis translation), up and down vertically (y axis translation), or in and out (z axis translation). We could also spin it around like it was the rotor of a helicopter (xz plane rotation), or spin it like a propeller of a plane (xy plane rotation), or spin it like we were rolling it along (yz plane rotation).

The way we spin something is to apply a force, preferably at a right angle, at a point as far away as possible from the pivot point, hinge, or fulcrum. One example is opening a door: we pull on the handle at a right angle to the frame of the door, and using the leverage as the width of the door, we rotate it open or closed. To be precise, leverage is the distance from the line of force to the pivot point, drawn perpendicularly. This is why long handled wrenches are much easier to use to loosen a nut, since the longer handle provides greater leverage. It is important to also note that the applied force must be at a right angle to the leverage to gain the maximum advantage, since if we pull along a line that passes through the pivot point, the result is no rotation at all. Anyone who has tried opening a door by pulling on the side of the door where the hinges are has learned this law of physics in an embarrassing way! Physicists have a name for a force applied through a leverage that causes rotation: torque.

For an object to be in static equilibrium, meaning it stays at rest in the reference frame and does not rotate, all six degrees of freedom must be satisfied in the following manner: all the horizontal (x) forces must add to zero, all the vertical (y) forces must add to zero, all the (z) forces must add to zero, all the torques in the xy plane must add to zero, all the torques in the yz plane must add to zero, and finally all the torques in the xz plane must add to zero. If these conditions are met, the object is either in static equilibrium or dynamic equilibrium (constant non-zero velocity). This analysis forms the foundation for thousands of applications in civil and mechanical engineering, as well as architectural design, such as buildings, bridges, and so on.

We have seen how mass is a primary factor in determining how an object translates through space in the form of linear acceleration. Given more mass, the same net force will produce less linear acceleration. But what about rotation? Does mass affect the angular acceleration of an object? Think of how we set a freely held bicycle wheel into rotation about its axis. We pull on the outside rim, tangent to a point on the outside edge of the wheel to maximize our leverage, and the wheel starts to spin. In brief, we apply an external torque to the wheel. At any given time, every point on the wheel has exactly the same angular velocity, assuming the wheel is rigid, but it should be evident that linear (sometimes referred as tangential) velocity varies greatly depending on where we "live" on the wheel. Since all points share the same angular velocity, points on the outside of the wheel must go through a much larger distance (circumference) in the same amount of time as inner points on the wheel. One of the reasons we like to launch our spacecraft from Florida is because of that fact: since it is close to the equator, the spacecraft already has a considerable linear velocity due to the Earth's rotation even as it sits "stationary" on the launch pad. In fact, a point on the equator is moving at around 1050 mph due to the spin of the Earth.

Anyone who has been on a carousel knows that it's a lot more fun on the outside rim then sitting at the center of the ride where one would just spin and have zero linear velocity through space! Therefore, as the bicycle wheel gains angular velocity from the applied torque, it must also be gaining linear or tangential velocity at any given point on the wheel. The two velocities are clearly directly related, and we will connect the two formally shortly in equation form. One can also see that if the wheel's angular velocity is changing, there must be an angular acceleration making that happen, and as a consequence the linear or tangential *acceleration* must be directly connected to its angular counterpart as well.

Now we are in a position to discuss what factors affect the angular acceleration of a wheel, given an applied torque. Mass is certainly involved, since it seems intuitive that a massive grinding wheel would be a lot harder to start rotating than a light bicycle wheel. But it gets a little more interesting the deeper we look into this phenomenon of rotational motion. Hold a heavy metal bar horizontally at its midpoint and spin it in a vertical circle about that center point. Then try to spin the bar the same way, but this time hold it at one end of the bar and use that point as the pivot point. You will find it much harder to spin the bar now, *even though its mass is clearly the same as it was before*. Even more striking, hold the bar horizontally and spin it about its long axis by rolling it in your hands: it is extremely easy to rotate it in this manner, and yet once again its mass has always remained the same. It follows that what influences the angular acceleration of something rotating, in addition to the amount of torque applied, is *how the mass is distributed in relation to the axis of rotation*. This resistance to a change in the rotation of a rigid object is called "rotational inertia" or "moment of inertia", and the formula used to calculate it depends on the distribution of the mass and often requires calculus to pin down since the mass may be spread throughout the object. Fortunately, nature doesn't require us to learn the math unless you are so inclined, but it can give us the results by doing the experiment! Place a golf ball, a hollow hoop, and a solid cylinder at the top of an inclined plane ramp, and release them from rest in a race down the ramp. Which one wins?

If you have done the experiment, making sure that all the objects continue to roll without slipping or sliding down the ramp (this is accomplished by ensuring there is friction on the ramp), then you know the result! It is also intriguing that the result will always be the same regardless of the relative masses or radii of the objects! The only factor that determines the winner is the way the mass is distributed in relation to the axis of rotation. Think about each of the objects: which one will be the most difficult to start rotating from rest? This is almost the same as asking which object has the greatest *intrinsic* rotational inertia. Now try a second experiment. If we could totally eliminate friction on the ramp, what do you predict would be the outcome of the race? I could state the results here, but I am not the authority: experiment provides the answer! I leave it to the reader to conduct the experiment. The frictionless ramp may be a challenge; perhaps you can do a "gedanken" or thought experiment and reach a conclusion in that situation.

When we tie all of this discussion about rotation together, we begin to see the correlation between linear and rotational systems. For linear motion, Newton gives us F_{net} = ma. Therefore, it seems reasonable that for rotational systems his second law of motion will instead read: net torque = moment of inertia times angular acceleration. In terms of a formula: $\tau_{net} = I\alpha$. Instead of force we have torque (τ); in the place of mass we have the distribution of mass or moment of inertia (I), and instead of linear acceleration we have angular acceleration (α). This is exactly the same law applied to rotating systems! There remains one item before tackling the fundamental force of gravity: what happens if there is zero angular acceleration?

If there is no net torque, then it must follow that there is zero angular acceleration. The wheel we have been discussing must have a constant angular velocity, which could be steady at zero or it could be some nonzero and unchanging angular velocity. Let us take the case of a wheel spinning at a constant rotational rate, since if it is not spinning then clearly the only thing left to do would be for it to sit there and do nothing, because if the wheel translates across the ground then it must also rotate (providing it isn't going through a massive skid by laying down a patch of rubber). Every point on the spinning wheel has the same angular speed, and outer points have greater linear or tangential speed than the inner points as we have discussed. Since we are taking the case of zero angular acceleration, then we must conclude that all points on the wheel have zero linear or tangential acceleration as well: there is no speeding up or slowing down going on. A significant problem emerges: a point on the rim of the wheel is not going straight, it is constantly changing direction, and that means it must be experiencing an acceleration since its velocity is changing even if its speed is not. Recall the steering wheel of a car is the third way we can accelerate, aside from the brake pedal and the gas pedal, and we can detect the sensation of that acceleration quite readily. But wait! We already said there is zero angular and linear acceleration in this example: does this mean there is yet another kind of acceleration for rotating objects? The answer is yes, there *must* be an acceleration for any object moving in a circle or curved path, otherwise it would keep going straight! Now perhaps you can see how this is going to lead into the heliocentric model and the force of gravitation between the planets and the Sun.

One could make the argument that Newton was obsessed with ancient alchemy, but he never did succeed in turning lead into gold. He was also passionate about figuring out the acceleration of something moving in a circle at constant speed. There are a couple of terms which require definition at this juncture in our analysis of rotational motion: centripetal means "center-seeking" and refers to anything pointing inward toward the center of a circle; centrifugal means "center-fleeing" and refers to anything pointing outward away from the center of a circle. It is the centrifugal piece that we will investigate quite deeply, because it is so often misused! Before doing that, the next six pages will be the only ones involving some advanced mathematics. Skip them if you like – they are only included because they show the beauty and power of mathematics and logical reasoning, and because they are directly connected to the law of gravity. The conceptual underpinnings of gravity, however, will be explained in addition to the math in the pages which follow. I leave it to the reader to make the call: study the math or move onward to page 170!

KINEMATICS: The Study of HOW Objects Move

Some textbooks provide mountains of formulas in kinematics, and I have found that to be needlessly confusing and redundant. In the final analysis, you only need two formulas:

For one-dimensional motion along the x-axis:

$\mathbf{V_{fx} = V_{ox} + a_x t}$ (Final velocity = initial velocity + the velocity gained/lost by accelerating)

This is just the definition of acceleration: note how we could rearrange the formula to get the acceleration "alone" on one side which would equal the rate of change in velocity! The subscript "fx" means final velocity in the x direction, and "ox" means original or initial x velocity. If you were to make a graph of this formula by plotting velocity on the y-axis and time on the x-axis, for a constant acceleration it would look like a line: note how this equation looks exactly like the form of a line: y = mx + b! *Using a direct analogy, can you see the slope "m" of the velocity vs time line must equal the acceleration "a" of the object?*

The above equation can be applied independently to motion in the y direction or in the z direction, just change the subscripts! No new formula is required because motion in each dimension of space is completely independent from any motion in the other two dimensions of space. The same holds true for the second equation of kinematics, which can be obtained from taking the area underneath the very same graph of velocity versus time used above. I will leave it as an exercise, but the area underneath the line is a trapezoid, which can be broken down into a rectangle and a triangle if you prefer. The X_o is the initial position of the object along the x-axis, which is often taken to be zero (at the origin) because it makes the math simpler and because space is isotropic: all points are essentially of the same value. Nature doesn't care at all where one chooses the zero point, as long as we remain consistent. Take the area of the trapezoid formed by the line in the velocity versus time graph and you will get:

$X = X_o + V_{ox}t + \frac{1}{2}(a_x)t^2$

These two formulas hold for any constant acceleration motion in one dimension. They can be used in the timing of traffic lights, accident investigations, rapid transit systems, runways, and thousands of other applications, as alluded to previously. The question then arises: what if the acceleration is *not* constant? This is where the work of Newton and Leibnitz enters in, and it is truly the stuff of genius: the branch of mathematics called calculus. Its power to analyze infinitesimal changes is extraordinary, and hence its applications are utilized in economics, statistics, and many other fields beyond physics. Constant acceleration, when applied to the speed of an automobile for example, would look like the following: every second, the speed increases by 5 mph. A table of values for the car's speed, entered every second, would look like 5, 10, 15, 20, 25, ... in mph. This is constant acceleration, assumed in a linear path for simplicity. Changing, in this case let's assume increasing, acceleration would look like this (as one example): 5, 10, 18, 28, 40, ...in mph. Note that in each one second interval, the change in velocity is getting larger! Rockets often will resemble this scenario, since they are losing mass in the form of fuel during powered flight. Calculus gives us a way of finding the position, velocity, and acceleration of any moving object *at any instant in time* provided we know at least one of the three variables (position, velocity, acceleration) at every time in the flight of the object and the function is continuous. Here is how it works.

Newton showed that finding the slope of a tangent to a curve, at any point on the curve, will give us the rate of change at that point. The slope of this tangent is called a derivative. Since the rate of change of velocity is acceleration, it then follows that if we know the function that gives the velocity of an object at any time, then the derivative of that function will be the slope of the graph (at any point) which is equal to the acceleration at any time. A couple of examples follow that illustrate this method in practice, without any intent of proving *why* it works! Suppose we have constant acceleration, and an object's velocity is given by $v = 20t$. This is a continuous function with a defined velocity at any time that is chosen. The results of calculus show that the derivative of a polynomial function works as follows: if $f(x) = bx^n$, then the derivative, denoted here by $f'(x) = nbx^{n-1}$. Don't worry about how that was figured out, unless of course you want to worry! The simple truth is, it works.

Now let's look at $v = 20t$. We have velocity "v" on the y axis and time "t" on the x axis. Here, the constant "b" = 20, and n = 1 (implicit since we don't write 1 as a power). The first derivative formula then becomes: $1 \times 20 \times t^0 = 20$ (since anything raised to the zero power = 1). This means the acceleration, which is the derivative of velocity versus time, is 20 units. But we knew that because (recall) it is the slope of the velocity versus time graph! This seems a bit trivial until we take the second example where the acceleration is not constant. Suppose $v = 15t^2$. When we plot that graph of velocity versus time, we do not get a line, we get a curve you probably recognize as a parabola. Now how does one find the slope of a curve, since it is changing all the time? Easy: use calculus! The acceleration at any time is equal to the derivative of the velocity function, which is, using the "recipe" for the first derivative of a polynomial: acceleration = a = 30t. Now they can ask you what the acceleration is at any time, say 4 seconds, and you plug in 4 seconds to get 120 meters per second squared as the acceleration at that instant in time.

How does one get the position as a function of time one might ask? Do the reverse, the "anti-derivative", which is called the integral and is equal to the area underneath the curve (which should sound familiar). In our first example where $v = 20t$, we ask what function would give us a derivative equal to 20t? Going backward, perhaps you can see that it would have to be $10t^2$ + a constant. We need to add the constant since we don't know if there was an initial starting position other than zero, and the derivative of any constant = zero (since it is constant!). To summarize, if we know the velocity of an object as a function of time, we can get its acceleration as a function of time by taking the first derivative of that function, and we can obtain the position of the object as a function of time by taking the integral of the velocity function. Perhaps you can also see that if we have the position as a function of time, then the first derivative of that function will give us the velocity, and the second derivative will give us the acceleration, both at any time we specify! This will prove exceedingly useful in all types of motion, including those influenced by the force of gravity.

Let's return to rotational motion kinematics by way of analogy. *The kinematic formulas are exactly the same in format, the only difference is the use of symbols!* Here are the rotational kinematic formulas – take note of how they "look" exactly the same as the linear ones:

$\omega_f = \omega_0 + \alpha t$

$\theta = \theta_0 + \omega_0 t + \frac{1}{2}(\alpha)t^2$

Instead of linear distance "x" we use angular distance "θ", which can be in degrees, revolutions, or radians (the preferred unit in rotation for mathematical reasons). Recall that 360 degrees equals one revolution which equals 2π radians, so as long as we remain consistent throughout the formula using the same chosen unit, we are copacetic. Instead of linear velocity "v" we use angular velocity "ω", in either degrees/sec, revs/sec, or radians/sec. Instead of linear acceleration "a" we use angular acceleration "α", in degrees/sec², revs/sec², or radians/sec². The process and format are absolutely identical!

Returning to rotational dynamics, we have seen how $F_{net} = ma$ (for linear motion) transposes to $\tau_{net} = I\alpha$ (for rotational motion). Instead of force we have torque, instead of mass we have the distribution of mass in relation to the axis of rotation, or moment of inertia, and instead of linear acceleration we have angular acceleration. Presented in this manner, there is no mystery to the analysis of rotational motion, as it is completely analogous to the linear motion equations. We now can return to the problem of uniform, constant speed motion in a circular path, and proceed to figure out what the "other" acceleration must be, since the angular and linear or tangential accelerations are both zero in this particular case, and we know we must be accelerating because the object is constantly changing its direction, hence its velocity.

Picture a circle of radius R, drawn on the XY Cartesian plane with its center located at the origin. An object is moving around on the circle at constant speed, orbiting the origin. A point on the circle could be specified by a coordinate (x,y) at any given time, but we also could specify the object's location by using what is termed "polar" coordinates, R and the angle θ it is making with the positive X-axis. The radius R is not changing of course, since it is a circle, but the angle θ is constantly changing since the object is moving around the circle. Therefore, we can use all our hard work on rotational kinematics and show that since there is zero angular acceleration, and we can always choose our original position to be zero, our kinematics formula reduces to **θ = ωt** (we can ignore the subscript for the angular speed since it is constant). This makes sense, since the angle we make will continually change as we move around the circle at constant speed.

Now place the object somewhere in the first quadrant, and let's define the coordinates x and y in terms of R and θ. Draw a line from the origin to the point on the circle where the object is, and make a right triangle by dropping an altitude perpendicularly down to the X-axis. By using basic trigonometry, can you see that x = R(cos θ) and y = R(sin θ)? Now substitute ωt in for θ, and we have x = R(cos ωt) and y = R(sin ωt). The final piece of the puzzle we need before finding the acceleration of this object, due to its constantly changing *direction* rather than speed, is to state the result, without proof here as it is a bit long (we will let the mathematicians worry about the details), that the derivative of the sine function is cosine, and the derivative of the cosine function is *negative* sine. Recall the first derivative of position versus time is velocity, and the second derivative is acceleration. Here are the results, doing the math:

X velocity = derivative of the function x = R(cos ωt) ➔ $v_x = -R\omega(\sin \omega t)$

(the extra "ω" is pulled out of the derivative because of something called the "chain rule")

Y velocity = derivative of the function y = R(sin ωt) ➔ $v_y = R\omega(\cos \omega t)$

X acceleration = derivative of -Rω(sin ωt) ➔ $a_x = -R\omega^2(\cos \omega t)$

Y acceleration = derivative of Rω(cos ωt) ➔ $a_y = -R\omega^2(\sin \omega t)$

Before assembling this result, take a look at what we have derived. The X acceleration is in the negative direction (toward the left), and the Y acceleration is in the negative direction (downward). *If we combine these two vectors, the net result is a vector pointing to the center of the circle. This means the acceleration of something moving in a circle at constant speed is a centripetal acceleration, pointing to the center of the circle!* Actually, this should make a lot of sense, since if we whirl a can around on a string, it is the tension in the string pulling inward on the can that makes it go around in orbit, and the acceleration is in the direction of this tension force. Now let's finish up the mathematics.

Since the accelerations a_x and a_y are components of the resulting acceleration to the center of the circle, we can find the magnitude of the centripetal acceleration by simply using the Pythagorean Theorem: $x^2 + y^2 = $ hypotenuse^2. Squaring both a_x and a_y components, we have:

[Centripetal acceleration]^2 = $[-R\omega^2(\cos \omega t)]^2 + [-R\omega^2(\sin \omega t)]^2 = R^2\omega^4[\cos^2 \omega t + \sin^2 \omega t]$

But we know the trig identity $\cos^2 \omega t + \sin^2 \omega t = 1$, so this simplifies greatly to the following:

Centripetal acceleration = $R\omega^2$ ... an amazingly simple result! Another way to express this result is to recognize that the linear speed v = ωR = 2πR/T, where T = the time to make one revolution

(2π radians). When plugged back into the above result, we obtain alternative expressions for the centripetal acceleration, herein referred to as a_c:

Centripetal acceleration = $a_c = v^2/R = 4\pi^2 R/T^2$

This result is of tremendous value for Newton, as he proceeds to develop his Universal Law of Gravitation. We will be using the result in the next section of the book. One begins to see how much work must be done to provide the foundation for formulating a law of physics with wide applicability!

For those of you who skipped the last six pages of mathematics, welcome back! We are now going to investigate the above result for an object moving with constant speed in a circle. Think of what you feel when driving a car around a corner, preferably at constant speed, but the discussion still pertains regardless. Clearly, the faster one goes around the curve, the more one "feels" the effect of changing direction. This is reflected in the formula for centripetal acceleration: notice that the speed "v" is squared in the numerator! For a given radius, doubling the speed of the car *quadruples* the amount of acceleration one feels. It is also an intuitive result that as we reduce the radius R of the curve, in other words make the curve tighter or sharper, the more we feel the acceleration – this agrees with the inverse relationship shown between a_c and R in the formula. Once again, I hope you see that formulas tell a story that we are often quite familiar with, just not in the formal language of mathematics. In both cases, speeding up and reducing the radius of the curve, we are causing our direction to be changing faster than it was before, which results in greater acceleration. This approach is how we design curves along highways and city roads. Assuming zero angular acceleration, Newton's Second Law gives us $F_{net} = ma = mv^2/R$; notice the mass "m" cancels out in the design of the curve. This is extremely lucky, since if it didn't, we would have to make multiple radii curves for different masses and it would be a mess! For a flat curve, the force which provides the necessary centripetal acceleration is the force of friction of the road on the wheels of the car. This makes it a poor design, not just in terms of drainage, but also in terms of the absolute dependence on friction to negotiate the curve safely. That is fine for nice, sunny days, but when the road is covered in snow and ice, we still need to be able to get around the curve without going off the road. To solve the problem of unpredictable (hence uncontrollable) weather conditions, we can bank the curve so that even with zero friction, we will make it around the curve safely, provided we keep our speed close to the design parameters (typically the speed limit or close to it).

Why does it feel as though we are getting pulled to the outside when we go around a curve? This is when people often invoke a centrifugal force, pushing out away from the center of the curve. Some analysis will reveal that this cannot be correct. First, suppose someone asks you what agent or cause is pushing you outward as you round the curve. You won't be able to identify any such cause, because there is none! Second, an observer watching you rounding the curve from a vantage point behind the car will see you actually trying to go straight or tangent to the curve, not outward, and straight-line motion is a product of inertia, not a force. The same holds for whirling a can around in a circle using a string: if we cut the string suddenly, the can does not fly outward away from the center, it flies off tangent to the circle at the instant the string is cut. We call this type of centrifugal force "pseudo" or "fictitious", since it only seems to arise in the accelerated reference frame and actually is just a manifestation of the law of inertia.

This is also the physics behind the origin of the famed "Coriolis Force", a pseudo force that arises because we live on an accelerated reference frame, the spinning Earth. If we launch a rocket from the equator and point it due north and subtract out any external factors such as wind, it will land a bit east of north because it has a (maximum) eastward velocity at the launch point due to the Earth's rotation. There is no real force pushing it eastward, just the spin of the Earth causing the effect. This is also how one can see how atmospheric circulations in the northern hemisphere will end up being opposite to those in the southern hemisphere, although actual weather patterns can be extremely complex due to a multitude of variables. It is a myth, of course, that toilets flush in opposite directions depending on which side of the equator one resides – the turbulence alone would wash out any (tiny) effects due to the Earth's rotation, and the design of the toilet itself dictates which way the water goes.

Imagine you are on a carousel ride and there is a person directly opposite you on the other side. As you both spin around at constant angular speed, your task is to throw a baseball to her so she will catch it. If you aim the baseball directly at her, you can see that it will miss the mark, and the faster you spin the more error will be incurred. To compensate for the accelerating reference frame, you aim a bit against the spin direction and hope you gauged it all correctly. Again, there is no real force involved here, just a very interesting rotating reference frame. The Foucault Pendulum is yet another great example that we live on a rotating platform called Earth – you may have seen one of these very long pendulums swinging back and forth at a science museum somewhere. When you return to view the pendulum hours later, the entire plane of its vibration has changed! It is now vibrating back and forth on a much different plane, and in fact by doing a little math we can figure out what latitude we live at based on how fast the plane of vibration varies. Of course, there is no force pushing this pendulum this way or that, it is the rotation of the Earth underneath the pendulum that only makes it appear so. The pendulum itself never varies from its original path, assuming no other external factors are present. This device provides great evidence that the Earth spins on its axis! At this point we have all the necessary tools in place to do a full exploration of the first fundamental force we experienced as humans, and yet we still don't fully understand: gravity.

GRAVITY

More than any other concept in physics, this idea called "gravity" led the way as an impetus for writing this book. I still find it remarkably mysterious why things fall to the Earth when dropped, and especially why it is that the acceleration rate of that fall is the same for all objects regardless of mass in the absence of air resistance. There are three wonderful books I want to bring to your attention regarding this topic, because in order to fully explore a concept as deeply interwoven into all of nature such as the phenomenon we call gravitation, many books need to be written and read! David Darling provides a broad historical sweep when investigating gravity through human history in his book, *Gravity's Arc*. He also has written a most profound book titled *Deep Time*. I encourage the reader to peruse both – you will find them to be enlightening! Lee Smolin takes us on a journey that is still ongoing, the search for the connections between quantum physics and the force of gravity, in his book *Quantum Gravity*. In this section, we are going to explore one possible path to Newton's historic formulation of the Universal Law of Gravitation, and the incredible volume and depth of the discoveries which have flowed from this law over the past three centuries. Without a doubt, this equation is a masterpiece on a par with Beethoven's *Fifth Symphony* or Michelangelo's *David;* it most definitely ranks as one of those few master equations that transformed human history.

Even with all the letters, papers, and books passed down through history, it is seldom readily apparent what the thought processes were whenever a great work of art, literature, music, science, athletic feat, or any such endeavor passes through the phases of creation. We know that Newton was familiar with the discoveries handed down by Copernicus, Kepler, and Galileo, and it is probably true that an apple falling in his garden caused him to at least ponder the idea that it falls to the Earth due to the same mechanism that causes the Earth to orbit the Sun. And yet, it was far from obvious that the physical laws which pertain to objects on Earth should also be universal in scope. Moreover, there was no fundamental, physical explanation for how objects could interact with each other when separated by huge expanses of space. It was evident to Newton, principally from the work of Galileo and Kepler, that an external, net force must be acting on the planets, otherwise they would travel in a linear path at constant speed. Heliocentric theory suggests that the agent causing the planets to deviate from this constant velocity motion must be the Sun. The process by which the Sun "communicated" this interaction with the planets over many millions of miles of empty space, deemed "action at a distance", often devolved into mystical nonsense of a distinctly unscientific nature. Newton was convinced that this mysterious force, translated from the Latin to "gravity", nonetheless had to exist.

For simplicity, we will assume circular orbits for the planets, leaving the most eccentric, elliptical paths for another day. The essential aspects of our analysis will remain intact as we uncover this universal law of nature along with Newton. From the previous section regarding the physics and mathematics of objects moving in a uniform circular path, we found the acceleration they experience is centripetal (toward the center of the circle) and equal to v^2/R. Using Newton's Second Law, $F_{net} = ma$, we can then set this net force called "gravity" acting on the planets equal to mv^2/R. The speed of an object moving in a circle is the same as the speed of any object moving uniformly: distance divided by time. Since the circumference of a circle is $2\pi R$, and the time for one revolution about the center is given as T, the period of motion, we can set the speed "v" equal to $2\pi R/T$. Now we substitute this expression for "v" into the force equation and we get the following expression: $F_{net} = F_{gravity} = m[(2\pi R/T)^2]/R = m[4\pi^2 R^2/T^2]/R = m4\pi^2 R/T^2$. Here, "m" is the mass of the planet in orbit around the Sun, "R" is the radius of orbit (assumed constant), and "T" is the period or time of one revolution about the Sun. Already Newton has truly stood on the shoulders of giants such as Galileo, Copernicus, Brahe, and many others, but now he connects directly with the work of Kepler – specifically his Third Law of Planetary Motion.

We came across the planetary laws that Kepler deduced from Tycho Brahe's exhaustive data over a hundred pages ago, on page 68, and now we reap the rewards. Kepler's Third Law was absolutely numerical and appeals to Kepler's mystical fascination with numbers. Recall he discovered that $K = R^3/T^2$, or the radius of orbit cubed divided by the period of revolution squared is exactly the same number (called Kepler's constant) no matter which planet one looks at! We can use this discovery to further explore this law of gravity by substituting for T^2 into the expression we have arrived at thus far: $F_{gravity} = m4\pi^2R/T^2$. Since $T^2 = R^3/K$, we now have the force of gravity as $F_{gravity} = m4\pi^2K/R^2$. It appears that the force of gravity, like the electrical force between two charges discovered by Coulomb more than a century later, is an inverse square law, meaning the force drops off as an inverse square of the distance between the two masses. Newton was a brilliant mathematician, and he knew before all of this that elliptical orbits must imply an inverse square law, so clearly he is onto something big here.

The careful reader may have noticed that I used the phrase "between two masses". You will notice that thus far, our force of gravity involves only one mass, the mass "m" of the orbiting planet. This is the point where Newton's remarkable genius kicks into high gear: he reasoned that Kepler's constant must involve the mass of the other object – the thing making the planets go around in orbit, namely the mass of the Sun! This makes perfect sense as a manifestation of his third law regarding action-reaction: if the Sun is pulling on the Earth, then the Earth must be pulling on the Sun. But he didn't stop there. He then reasoned that *all mass attracts other mass in the universe*, without exception. Can you see the apple falling to the Earth now, and all sorts of insights flashing through the air? The Sun and the planets is just one example of a universal interaction, and so the $4\pi^2 K$ term could be set equal to some constant times the other mass! In one fell swoop Newton made his law of gravity universal, not just applicable to the planets, moons, stars, and galaxies. In equation form, his masterpiece materialized as $\mathbf{F_{gravity} = GMm/R^2}$. The constant "G" is labeled the universal constant for gravitation, and it holds for any two objects in the universe. We will get to that in a moment, but for now note that the force of gravity involves the mass of two objects (M and m), and the distance (R) between them squared! Newton did not know the value of this constant G, but he made some extraordinary guesses based on the average density of the Earth (primarily rocks) and its known size. He also knew that G had to be a very small number, since "normal" masses such as human beings, books, horses, even houses don't appear to be attracted to each other, at least as far as could be measured at the time. The first experimental value for G was to be found by Henry Cavendish, nearly one hundred years later in 1798, long after Newton died.

Had the Nobel Prize been awarded back around 1800, I am convinced Cavendish would have earned the esteemed prize with his most ingenious design – an experimental setup that enabled him to find G to within 1% of today's accepted value. Here is the essence of what he did, and I include this description because once we have this value for G, the things we can deduce from the Universal Law of Gravity proposed by Newton are fascinating and far-reaching. As you have probably come to expect, several others before Cavendish had thought of this experiment, and a few had even progressed to the point where they had designed an apparatus to measure the gravitational attraction between two masses of ordinary size. The basic features of this setup are as follows. A wooden bar is hung from a fiber or thread so that it is perfectly balanced at its midpoint, resting horizontally and parallel to the floor. At either end of the metal bar, a small lead sphere is attached or suspended, keeping the entire apparatus in equilibrium by making sure the lead spheres have identical masses. The next step is to bring two much larger lead spheres close to but not touching the small spheres, with one large sphere being on the side closest to an observer, and the other large sphere placed on the back side of the wooden bar. The idea behind this simple arrangement is that the gravitational attraction between the large and small spheres should cause the wooden bar to twist ever so slightly. Using the formula for the Universal Law of Gravitation (herein referred to as ULG), $\mathbf{F_{gravity} = GMm/R^2}$, we can see that the masses are easily measured in the lab, as well as the distance R between their centers. As a not inconsequential aside, Newton had already solved the problem of the distribution of mass in a solid sphere, since one might argue about which points should be used to measure that distance. He showed, using the same calculus principles he invented, that one could consider the entire mass of the solid sphere to be at a single point located at its center, thus eliminating that hurdle in terms of defining the separation distance between masses (clearly the Earth and the Sun factored into this issue). Now that we have the masses, M and m, and the distance R, the only obstacle remaining in order to determine the value for G is to measure the force F between the masses. That was not as simple as it may sound, since the force of gravity is extremely weak unless the masses involved are huge, like planets and stars.

Cavendish was able to measure the force by timing how long it took the torsion fiber to swing back and forth through a large number of cycles. Every child on a swing knows how much fun it is to twist the swing around and then rotate back and forth for a while – the principle of a torsion fiber is essentially the same. Knowing the rotational period of the torsion fiber then tells us the torsional constant, which in turn gives us the necessary torque and thus force we seek.

Cavendish had to be very careful to damp out any external disturbances near the apparatus, since the amount of "twisting" was so small as to be almost imperceptible, and any nearby rumbling or movement could set the apparatus into motion, easily washing out the gravitational effects. He appears to have conducted the experiment in a shed away from any residential setting, and fortunately during this time in history there were far fewer disturbances such as trucks, cars, and other heavy machinery. The present day value for G, determined in much the same manner as it was hundreds of years ago, is 6.673×10^{-11} Newton x meter^2/kilogram^2. Since a "Newton" is a unit of force that is kilograms x meters/second^2, this constant reduces to the three basic concepts of space, time, and mass. It is worth repeating: Cavendish was able to nail this down to within 1% of the value for G we use today! This was not the only contribution made by Cavendish in the progress of science, but it ranks as one of the most important to be sure.

As expected, the constant is an extremely small number, reflecting the fact that gravity is a very weak force for ordinary masses. There are a couple of points to emphasize here, and then we will proceed to the amazing results which follow from knowing G. First, it is assumed that the value for G is *universal* in the literal sense: it applies to all masses everywhere in the universe. Second, it is assumed that this value is constant through time and has never changed. This (second) assumption is under scrutiny by many physicists these days. Both assumptions have held up well in every experiment we have done thus far, although of course we haven't been able to do this experiment in a distant galaxy as yet! Using the ULG and the known weight of any object on Earth, *we can now figure out the mass of the Earth*! Indeed, Cavendish often referred to his experiment as weighing the Earth. Since the weight of a person is mass multiplied by the acceleration of gravity at the surface, and we can measure the acceleration of gravity by dropping an object and using simple kinematics (or by timing a pendulum's swing), we have the only unknown as the big M in the formula for the ULG. It looks like this in mathematical form: $mg = GMm/R^2$ ... so "m" cancels out (all mass falls at the same rate of acceleration!), G is now known, and R is just the radius of the Earth in this case, since we stand at its surface, and it has been well known for thousands of years. The only thing left to do is to plug in the numbers and solve for M, the mass of the Earth. When we do that, we find $M_{Earth} = 5.98 \times 10^{24}$ kilograms. Think on that for a moment. The human mind has now figured out the mass of a planet, something that could never be done directly since there is no way to actually place the Earth on a scale!

Recall that Edmund Halley was instrumental in getting Newton's *Principia* in finished book form, and he also helped with the enterprise financially. Halley is best known for his prediction, based on the ULG, that "his" comet would return every 76 years in its eccentric orbit around the Sun, but like most scientists of his day he was a generalist in the sense that his contributions spanned a large array of what are now separate scientific fields. Halley's Comet was an object of abject terror until it was rendered completely predictable by the ULG, and yet superstitions and irrational fears persisted about it even into the 20th century. It is a valid historical question as to whether Newton would have ever published his work without the assistance of Halley, especially because his original calculations regarding the period of the Moon about the Earth were off by a considerable margin. Newton did not need G to make this prediction of the Moon's period, because the distance to the Moon was known to be about 60 Earth radii, and using the inverse square law built into gravity, he was able to deduce its acceleration around the Earth, hence the amount of time it should take for one complete revolution. Unfortunately, he initially used an inaccurate value for the radius of the Earth and that is why the prediction for the Moon's period was far enough off for Newton to shelve his work for years. When Robert Hooke and Halley paid Newton a visit, and the correct value for the Earth's radius was found, the ULG was revived and eventually published. Though Hooke became an intense intellectual enemy over time, claiming that the ideas inherent in the ULG were really his and not those of Newton, it is important to recognize the critical need for collaboration among scientists if we are to progress. Newton was fortunate, despite his frequent battles with several contemporaries, that he also had a steadfast friend in Halley. He died a famous and revered man, and perhaps remains as the greatest physicist in human history.

The story of the ULG only gets better for the next 200 years. Since gravity pulls objects toward the center of planets, stars, and so forth, it is clearly a centripetal force. *This means that for any object orbiting a central mass, we can set the ULG equal to the centripetal force acting on it!* Again, we assume a circular orbit for simplicity, and thus the mathematical formulation looks like this: $GMm/R^2 = mv^2/R$. The small "m" here is the orbiting body, whether it be a moon, planet, star, or even an entire solar system. One can see it cancels out from both sides of the equation. The speed "v" depends only on the radius R of orbit and the time T for one complete revolution (recall $v = 2\pi R/T$), and G is now known thanks to Cavendish. This means we can calculate the mass of any central mass M if we can figure out the radius of orbit and the period of revolution for the orbiting body. This leads to some astounding predictive power.

Let us start with the Earth as it moves around the Sun. We know the radius of orbit, and of course we know it takes one year to complete a revolution. Plugging in the numbers, we can find the mass of our Sun, which turns out to be about 2×10^{30} kilograms. We could also check the mass of the Earth by looking at the Moon's orbital radius and period, and we get a result the same as when we used freefall and our weight "mg". We can now find the mass of any planet with moons going around it, since we can visually measure the period T and deduce its radius of orbit R around the planet with some fairly simple geometry. Thus, the mass of Mars and Jupiter can be calculated, as well as any other planet with observable moons. (The masses of Mercury and Venus, which have no moons, can be deduced by other means involving what is termed perturbation theory – the influence of other planets of known mass on their orbits.) This in turn means we can find the acceleration of gravity on these planets, knowing the radius of the planet itself, by just working backward and finding "g". Think of how significant that knowledge has become as we send probes to Mars and actually land them on its surface to run thousands of tests and gather samples. We could never have done this without the ULG. The amount of information we can now gain about our solar system is immense without ever having to actually go there, although I think we should still try to go because one never knows what we might find, and also because it is good for the human spirit to explore!

We save the most remarkable for last: the prediction of the mass of our galaxy. Since we have figured out our distance from the galactic center through the luminosity of stars, parallax, and some ingenious methodology, the only remaining challenge is to estimate the time it takes for our entire solar system to make one revolution around the center of the galaxy. The latter can be done by measuring how much time it takes for us to move through a small angle relative to the galactic core, then extrapolate to 360 degrees. In this scenario, the rotating object "m" is essentially the Sun since it contains nearly all the mass of our solar system, and the central object "M" is the mass of the Milky Way's galactic core, which also contains most of the mass of the galaxy. Measurements have found the radius of orbit R to be about 25,000 light years (recall our trip through the universe at the speed of light!), which we can convert into meters, and the period of revolution T is about 230 million years, which we can convert into seconds. Plug in the numbers and we get the mass M to be about 1.5×10^{41} kilograms. Since the mass of our Sun is about 2×10^{30} kilograms and it is an average, main sequence star, we can deduce that there are billions of stars at or near the center of the galaxy. Actually, we know there is an enormously massive black hole at the center of our galaxy as well, and that is a subject of ongoing and intense research today.

Every satellite we put into orbit depends on the ULG for proper altitude, orbital speed, and orbital period. Geosynchronous satellites, made to orbit in such a manner as to remain stationary over a defined location on Earth (useful for monitoring weather and many other localized parameters) by "matching" or synching with the spin of the Earth, must be located at a specific altitude of orbit to make this happen, again calculated by using the ULG. The ocean tides, in fact all tides on Earth, can be calculated from the difference in pull of the Moon, and to a lesser extent the Sun, from the near to the far sides of the planet or body of water. Though actual tidal movements are extremely complex due to a synergy of multiple variables at play, the primary factor is the proximity of the Moon to the Earth and the dependence of gravity on distance. Even though the Sun pulls on the Earth with greater force than the Moon does (which is why we orbit the Sun and not the Moon), it is so far away that the *difference* in pull from the near to the far side of the Earth is quite small, producing a stretching or tidal bulge that is considerably smaller than that of the Moon. Since the Moon is much closer to the Earth, there is a substantial *difference* in pull from the near to the far side of the Earth, so in a sense the Earth stretches out like taffy due to this imbalance of force across the planet.

And so the ULG reigns supreme for 200 years, and all is well with the Newtonian universe during that time. The nagging problem remains, however, as to the nature of gravitation: the *cause* or the mechanism for action at a distance is unanswered and a deep mystery. Many fanciful ideas are set forth to explain what makes the Earth and other planets swing around the Sun as though they were tethered by an invisible lasso. None of them make sense; none of them work. It is noticed that the orbit of Mercury is deviating a little from the predictions of the ULG, enough of an effect to elicit consternation about the accuracy of the law. These are distant rumblings at first – small murmurs of discontent on the horizon despite the overwhelming successes of Newtonian physics in general and the ULG in particular. The latter had even been used to predict the existence of a planet, Neptune, based on some small deviations observed in the orbit of another planet, Uranus. Galileo had actually spotted Neptune many years prior to this (through a telescope since it is not visible to the unaided eye), but mistook it for a star. Yet based on Newton's Laws and the ULG, it was predicted that another planet must be out there, and shortly thereafter Neptune was (re)-discovered. It would take some kind of fantastic revolution to supplant such a successful theory of gravity, one that had predicted so many things successfully over many years.

That revolution came in the guise of a most unlikely person, originally unconnected to mainstream physics and laboring away as an obscure patent clerk while pondering what the world would look like while traveling alongside a beam of light. Enter Albert Einstein - the world of physics was about to be turned on its head and vastly expanded in ways difficult, if not impossible, to imagine at the time. With his series of seminal papers submitted in 1905, Einstein catapulted himself from anonymity to front page physics news in a very short span of time. Recall in the Special Theory of Relativity that constant velocity reference frames were considered equivalent. This marvelous insight traces all the way back to Galileo. In practice, it means that no experiment can be conducted to distinguish one constant velocity from another: all physical laws, hence natural processes, proceed in exactly the same manner. As one example and to review, if we are in a rocket moving at constant velocity and we flip a coin, hang a pendulum vertically, or *conduct any other experiment,* the results will be exactly the same as if we were at rest or moving at some other constant velocity: the coin will land back down in my hand, the pendulum will hang vertically and swing in the exact same way. Add to this idea the experimental fact that the speed of light is the same (or absolute) to all observers, and we swiftly were on a journey where the measurements of space and time were relative – dependent on the reference frame of the observer. Einstein was not satisfied with this, however, because like most physicists he held a deep conviction regarding the symmetries found in nature. To wit, he posed the question that if all constant velocity reference frames are equivalent, then perhaps *all accelerated reference frames share an equivalence also, and this of course includes living in a gravitational field*!

Imagine you and your friend occupy two rockets. In your rocket, you are blasting through space with an acceleration of exactly one "g", just as we did on our trip through the universe. Therefore, you will feel very comfortable since you will have the same sensation as your normal weight while resting on the surface of the Earth. Your friend's rocket has not yet blasted off, and it rests on the launch pad at the Earth's surface. She, too, feels the familiar one "g" of gravity known as her "true" weight. Both space capsules are fully enclosed for our purposes, so we cannot see outside or use any external reference point to discern relative motion. How can we distinguish between the acceleration of your rocket and the gravity your friend experiences? What experiment can be conducted that will tell us which is an inertial acceleration and which is "just" gravity? The surprising answer given by Einstein is: you can't – there is no experiment that can distinguish between the accelerated reference frames, and therefore there is no difference between inertial mass and gravitational mass. Once again, Galileo was onto something big when he showed that all objects fall at the same rate of acceleration, regardless of mass. Though it may not seem like a revolutionary idea at first, if one pursues this equivalence, some extraordinary predictions rise to the surface, encapsulated by Einstein's General Theory of Relativity (GTR).

If being in a gravitational field is identical to being in an accelerated reference frame with the same value as "g", then all experimental results in either frame should agree as well. Suppose we aim your rocket vertically upward while it is accelerating through space at one "g", and someone fires a laser perpendicularly through the rocket as it accelerates upward. To an observer in the rocket, namely you, the laser beam must exit the rocket *below* the point of its entry, since the reference frame is accelerating upward while the beam is traversing the span of the rocket. To be sure, the deviation in your reference frame from a straight path for the laser is minuscule since the speed of light is so huge, but there is no doubt that it must curve slightly (as a parabola), just as throwing a baseball horizontally in the gravitational field of the Earth causes the ball to arc downward. The conclusion is inescapable due to the equivalence principle: *this means light itself must bend or curve in a gravitational field*! Einstein's concept of gravity in the General Theory of Relativity is that mass itself causes spacetime to curve, much like dropping a bowling ball on a flat mattress would cause a "dent" whose curvature objects nearby would be constrained to follow if set into some type of tangential motion. This is a most remarkable concept: gravity as the geometry of spacetime, and the more massive an object, the greater the curvature it creates. Since everything we know exists in this four-dimensional spacetime, including light, all of it must follow the curvature it resides in!

At this point, if Newton were alive he might jump out of his seat and loudly proclaim some alternatives. How can light, which we presume has no mass, be affected by gravity? The very essence of the ULG posits that gravity is caused by the influence of one mass on another. What is going on here? In fact, you may recall that Newton adhered to the "corpuscular" theory of light, and with such a scheme he proposed in his book on *Optics* that light *does* have a (small) mass and thus should be affected by massive objects such as stars. One can calculate the amount of deflection using this concept and the ULG. It turns out that the Newtonian prediction for the bending of starlight was about half the amount of angular deviation predicted by Einstein's GTR. We now have three competing models: one theory (GTR) predicts that light will bend or curve in a gravitational field, another (ULG) holds that if light has mass then the curvature of light near a massive object will be roughly half as much as predicted by the GTR, and finally the ULG would predict no deflection at all if in fact light has no mass. Once again, the power of scientific methodology to establish truth illuminates the path to enlightenment. Nature neither asks for nor cares about which theory we choose to believe, because if we are humble and wise enough to listen and observe, it is *evidence* that will guide our way up the mountain.

If one sees the Creator or the Great Spirit or God, or whatever the terms used for such a belief as permeating throughout all of nature, then it seems to me to be the height of arrogance to ignore the patterns she provides for us along the way in favor of personal beliefs. It is like saying a tiny bundle of atoms, within an immense universe of atoms and much more, can better explain all the mysteries of the cosmos than what the whole cosmos is trying to reveal through actual data. And in my view, the answers bring us closer to that Creator, and closer to the great beauty in nature that is most assuredly spiritual as well as scientific. Why we insist on this dichotomy between science and religion, or that one spiritual path condemns all others to a lesser existence, is supremely unfortunate and often tragic in terms of the human consequences. I have been reading *The Meaning of it All* by Feynman; he writes about some of this in his book. I encourage the reader to explore many such avenues of approach through life. Anyway, I digress. The point is the same as always: when it comes to competing theories, experiment provides the answer!

In the year 1919, two expeditions set out to measure the deflection of starlight passing by our Sun: one went to Brazil and the other to Africa. Naturally these measurements had to be done during a solar eclipse, hence the timing, weather, and many other factors (including the tumult at the end of World War I) had to coalesce in just the right way to make these observations possible. There are several wonderful accounts of these expeditions done by Sir Arthur Eddington (Africa) and Frank Dyson (Brazil), and I leave the reader to explore them in detail. For our purposes, let's skip to the results: both expeditions proved, despite some initial controversy, that starlight does in fact bend when passing by a massive object like our Sun, and the amount of the deflection agreed with Einstein's GTR, not Newton's ULG.

Einstein had achieved exceptional notoriety within the scientific community by this time, but his successful prediction of the bending of starlight was trumpeted throughout the world and he instantly became famous throughout every walk of life for all his remaining years, much to his chagrin. The best account I have read that takes the reader through Einstein's remarkable life is titled *Einstein: His Life and Universe*, by Walter Isaacson. It is a terrific compilation of primary source notes and letters, and it conveys a full appreciation for the creative genius and imagination of this extraordinary man.

We have derived Newton's ULG with a modicum of algebra, though it is important to point out that calculus was required along the way. The equations of GTR are far more sophisticated and complex mathematically. While time dilation and length contraction can be deduced from Einstein's two postulates of Special Relativity (equivalence of constant velocity reference frames and the fact that the speed of light is the same to all observers), the path to the governing equations of GTR is very high terrain even for the most advanced mathematicians. There are dozens of ways these equations are written, but I will present just one of them, and it is the one I can find the proper font and symbols for:

$$G_{\mu\nu} + g_{\mu\nu}\Lambda = \frac{8\pi G}{c^4} T_{\mu\nu}$$

Here, the left side of the equation describes how mass warps or shapes the geometry of space-time, while the right side explains how that geometry manifests itself in the motion of objects within the curvature produced by the presence of that mass. One might recognize G as Newton's gravitational constant, determined by Cavendish, and c as the speed of light. The other symbols represent what are called tensors or scalars or a cosmological constant. That last one, the cosmological constant, is a bit of extraordinary history that is a subject of much research today – it's the symbol shaped like a mountain that immediately follows $g_{\mu\nu}$. I leave it to the reader to further investigate what all of the symbols are in practice, but we will return to this cosmological constant shortly. A full exploration of the field equations is the subject of an entire course, and perhaps an entire life. Instead, I want to finish this section on the first, and the weakest, of the four fundamental forces we label gravity by presenting more experimental evidence that GTR works. All of this evidence is remarkable, but the last piece is positively stupendous.

We should start by reiterating that Newton's ULG has not been discarded in any sense of the word: we still use it to navigate to the Moon, place satellites in orbit, and for many other such endeavors. The mathematics is accessible to many, and the theory works in most scenarios. Einstein's GTR is a superior model in that it predicts all of what Newton's ULG does, and then goes on to successfully predict much more. The mathematics becomes prohibitive for many people, but all physicists would agree it is one of the most brilliant achievements of the human mind in our history. Let's start with the phenomenon exhibited in Mercury's orbit about the Sun. The ULG could not explain or predict the correct amount of what is termed the precession of the perihelion of Mercury. This means the plane of the elliptical orbit Mercury exhibits around the Sun actually (slowly) rotates about the Sun over time. The GTR successfully predicted and verified the rate of this precession, one of the first triumphs of the theory. Let's move on to some more recent discoveries.

GTR predicts that the greater the mass, the greater the curvature in spacetime it produces. This greater distortion in the fabric of spacetime in turn causes time to slow down. Therefore, we would expect time to tick out slower in stronger gravitational fields. The effect is small, but with our extremely precise cesium clocks discussed many pages ago, we can measure any predicted time differential in a number of ways. First, place a cesium clock at the base of a mountain where gravity is a bit stronger (the curvature of spacetime is more pronounced) and another identical cesium clock at the top of the mountain. Leave the clocks there for a few days, then bring them back together and compare the amount of time that has elapsed on each clock. It turns out the clock at the bottom of the mountain ticks out time slower by exactly the amount predicted using the GTR. Now suppose we do the same experiment, but this time we place one of the clocks on satellites orbiting the Earth. The effect should be even more pronounced in this instance because gravity or spacetime curvature is much less at orbital altitudes. Again, the GTR works perfectly, and here is the clincher: since our GPS navigational systems work through satellite technology and extremely precise synchronization of clocks on Earth with those in the satellites, unless we corrected for this time differential predicted by the GTR, our GPS systems we use to navigate around the globe would be completely useless!

We save the best bit of evidence for last: the prediction from GTR that gravitational "waves" should be present in the universe if extremely massive objects come in close contact with one another. A rough analogy would be to think of the universe as this vast mattress with bowling balls causing dents (spacetime curvatures) throughout, and thus if the bowling balls somehow collided or came in close proximity with one another, they might send ripples across the fabric of spacetime that were large enough to be detected elsewhere at a later time. This is the basic idea behind the new technology called LIGO: Laser Interferometry Gravitational-wave Observatory. We have two such observatories in the United States, one in Louisiana and the other in Washington state. There are others around the world in various stages of operation, one in Italy, another in Germany, and another under construction in Japan. Here is how they work to detect this possibility of gravitational waves traveling throughout the universe.

We are going to strip away many of the intricacies of the LIGO apparatus, and drill down to the basics of how it might detect gravitational waves through spacetime. This may seem like fanciful science fiction, but we now know it is quite real. Laser light is split so that it runs along two "arms" or tunnels that are perpendicular to each other and *exactly* the same length. In the case of LIGO, the arms are each typically over 2 miles long. When the light beams return to the starting point, they combine in ways similar to the double-slit experiment done by Thomas Young way back in 1800: they interfere either constructively or destructively. The trick is to make this apparatus so sensitive that any "shift" in this interference pattern is detectable. The shift in this pattern would be produced by the path lengths traveled by the laser beams altering ever so slightly. Obviously, tremendous care must be taken to eliminate any external disturbances that would cause the arms to vibrate, thereby moving or shifting the location of the mirrors used to split and reflect the light. This is no simple task and is one of the most challenging aspects of LIGO design. Here is the engineering piece that is truly amazing: LIGO is built to detect a change in distance between the mirrors of 1/10,000 the width of a proton, or 10^{-19} meters. That degree of precision is almost unfathomable, but it has been accomplished!

If a gravitational wave were to pass over the Earth, the "ripple" would cause one arm of the interferometer to stretch a little while simultaneously the other arm would compress, and then the cycle would reverse so that the arms would compress and stretch the other way, and so forth as long as the gravitational wave was passing through. Literally, the gravitational wave is stretching and compressing the very fabric of spacetime we live in – it's like living on a raft on a lake and a distant boat we cannot see or hear travels by, and minutes later our raft starts to bob up and down a little due to the waves traveling by us. The LIGO system vibrates like a spring momentarily, and the changing distances in the arms will be detected by the oscillating interference pattern seen at the detector. Once the observatories are online, the waiting game begins in earnest.

We did not have to wait very long. On September 14, 2015, the first gravitational waves ever detected were recorded at the LIGO facility in Louisiana, and about 7 milliseconds later the same signal washed over the facility in Washington. Einstein predicted the existence of these waves over one hundred years ago, but the technology was not available until very recently to detect such a weak signal. He also predicted that the waves would propagate at the speed of light, and so the 7 millisecond delay, when combined with the known distance between the facilities, corroborated that the signal was traveling across the planet at the speed of light. The frequency of the detected wave started at around 35 Hertz and rose to a level of 250 Hz, and since both frequencies are well within the range of human hearing, we can actually "hear" the wave passing by as well! This was the infamous "chirp" as it was called and replayed billions of times across the planet. What was the cosmological event that caused this "chirp" that was detected? The signal received provided many clues, and the results are nothing short of spectacular.

Two black holes approach one another, and begin orbiting about a common center of mass. The closer they get to one another, the faster they orbit around, much like how a tetherball spins around the pole faster and faster as it approaches the pole, except in this case it is two massive objects swirling about each other. Since black holes are supermassive collapsed stars, the curvatures in spacetime are immense here, and as they spin about each other with ever increasing frequency, we can see why the signal we detected likewise increased in frequency until they merged. The strength and frequency of the signal gives us a ballpark estimate as to their masses and their proximity to one another at various times, and it also tells us that this cosmic dance occurred over one billion years ago, creating a gravitational wave that traveled at the speed of light for over one billion years until its very much reduced amplitude washed over the Earth in 2015.

If that doesn't blow one's mind then nothing will. *We are able to observe the residual effect of two black holes merging billions of years ago, and all of this was predicted by Einstein in the early part of the 20th century.* The location of this primordial dance was harder to pin down due to the fact that only these two LIGO systems were in operation, but it has been narrowed down to a relatively small slice of sky in the direction of the Magellanic Clouds, but much, much farther away. Dozens of gravitational waves have been detected since that time by the LIGO system, and I find each one to be an absolutely stunning achievement of experimental prowess as well as theoretical genius.

Now we return to that cosmological constant Einstein inserted into his GTR field equations, a maneuver he later dubbed "the greatest blunder of my life". He included it in the equations because if he didn't, the theory predicted the universe was not static and was subject to expansion or contraction. One must remember at this time in history, the only galaxy we knew about was the Milky Way, and it was widely assumed that the entire universe was comprised of this one galaxy, floating around in space. Einstein *believed* that the universe should be static and essentially unchanging. By inserting the cosmological constant he "forced" the equations to allow for static conditions, even though there was no evidence or real scientific footing to do so. Only a decade later Hubble and others discovered there were actually billions of other galaxies, and even more revolutionary, the fabric of spacetime was stretching: the universe was indeed expanding! Einstein must have been quite angry with himself for this, because he could have been the first person to predict that the universe was expanding if he had let the equations be, rather than introduce a "fudge factor". And yet, the story doesn't end there. Now that we have learned that the expansion of the universe is *accelerating*, the cosmological constant might be necessary after all, and perhaps it has something to do with that dark energy discussed previously. This is a topic of much speculation and research today. Nobody has the answers yet.

And so this remarkable journey to understand the fundamental force of gravity continues, as it has for thousands of years, since it was the first and most ubiquitous phenomenon known to humans. One might argue that we should not call it a force at all, since the modern conception is that it arises from the geometry inherent in the curvature of spacetime produced by mass. Is Einstein's hugely successful GTR the final resolution to this enigma labeled "gravity" by Newton, so many years ago? The answer, predictably, is...not yet. The GTR remains incompatible with the precepts of quantum physics. Fundamental forces are proposed to be communicated through the interaction of "exchange particles", and in the case of gravity these particles, called gravitons, have yet to be detected. While inside black holes all physical laws appear to break down, outside black holes on a cosmological scale the GTR works exceedingly well, until the equations are applied on the atomic level. When that occurs, predictions start to unravel as infinities pop up and nonsensical results flourish. When we use GTR to try to explain the evolution of the universe close to the time of the Big Bang, it doesn't work. Something is missing, and the search goes on in the attempt to weld the two highly successful fields of physics (in their own realms) together into one unified whole.

STRONG NUCLEAR FORCE

Since we started with the weakest of the four fundamental interactions that govern the universe (gravity), it seems reasonable to jump to the strongest of the four now. We've had direct experience with gravity for many thousands of years, but the nuclear forces (strong and weak) have only been discovered and analyzed over the past 100 years or so. When one jumps into the field of subatomic particle physics, it is very easy to get completely swamped with all the vocabulary used to identify various particle types: we have muons, bosons, fermions, hadrons, mesons, pions, baryons, and many more. Some of these words represent whole categories of particles, others are subsets within those categories. I was fortunate enough to be involved with an electron scattering research group that was investigating the structure of the simplest nucleus we know of and which involves the more familiar protons and neutrons: the deuteron. A deuteron is the nucleus of an isotope of hydrogen called deuterium, and its constituents are exactly one proton and one neutron. Collectively, protons and neutrons are classified as nucleons. I did a lot of calculations for this research group, which meant I became quite familiar with complex integral tables in the CRC Handbook of Chemistry and Physics! I will start our exploration of the strong nuclear force with the element hydrogen, because that is the simplest and most logical point of departure for this leg of the journey.

Recall the work of Rutherford et al, done in the early part of the 20th century, in which the atom was discovered to have a positively charged, extremely dense nucleus by scattering positively charged alpha particles (helium nuclei) off a sample of thin, gold foil. Whereas the atom had been found to have a size of about 10^{-10} meters (from one of Einstein's "year of magic" papers in 1905), called an Angstrom, the nucleus was determined to be about 10^{-15} meters in size, called a Fermi, from these scattering experiments being done. This nuclear unit is in honor of Enrico Fermi, who made significant contributions in nuclear physics, including the crucial step of a sustained chain reaction in the fission process – more on that later. Thus, the proton entered center stage, and was soon to be accompanied by the other nuclear constituent, the neutron (recall this nuclear component was discovered by Chadwick just after 1930). We look first at hydrogen with a lone proton in its nucleus.

Current theory predicts, from something termed the Standard Model, that protons are composed of three quarks, a word drawn from *Finnegan's Wake*, by James Joyce. There have been six types of quarks identified to date, and they are called Up, Down, Strange, Charm, Bottom, and Top. For the proton, this internal structure is deemed to be two Up quarks, each with a charge of +2/3 of the elementary charge, and one Down quark, with a charge of -1/3 of the elementary charge. Therefore, the total charge of the proton is 2/3 + 2/3 + -1/3 = +1 elementary charge, the same as the charge on an electron, except it is positive instead of negative like the electron. (Again, the number of protons and electrons in any atom is typically the same so that the atom is usually electrically neutral.) Before going any further, it is important to note that at the time of this writing, *we have never observed an isolated quark or fractional elementary charge experimentally*. We use this model to explain things we *can* observe experimentally, and which are predicted by using this theory as a foundation. Though we know we can, and have, split the atom – more precisely we can split the nuclei of certain atoms – it appears we cannot split the proton into its three individual quarks. The question is: why not?

One aspect of the strong nuclear force appears to be that quarks are permanently confined inside a proton. We visualize this confinement by invoking a property called "color charge" which has nothing to do with actual colors or actual electric charge, but is useful only as a descriptor for a property of quarks in general. In this scenario, quarks stick together through the exchange of gluons, appropriately named because they "glue" the quarks together. It appears at the moment that no matter what we do in our attempt to split the proton apart into three distinct quarks with perhaps a huge influx of energy, the strong nuclear force finds a way to retain zero net color charge in such a manner as to instantly snap the proton back to its original composition. While I am loathe to assume that splitting a proton is impossible, we can state for a certainty that it hasn't been done yet, and the prospects for doing so seem unreachable.

Some models of the extremely early stages of the inflationary big bang universe picture a sea of quarks and electrons milling about until at some point with sufficient cooling the quarks become permanently confined within protons and neutrons. There is some research and speculation, however, that perhaps quarks are made up of things called preons, and these exceedingly dense particles might have somehow escaped, accounting for some of that dark matter we discussed many pages ago. One might begin to see there is a significant amount of speculation, guesswork, and imagination involved in the determination of the ultimate structure of matter. This includes the much publicized conjecture of a "theory of everything", called string theory, that may have lost some of its original steam in terms of progress and adherents according to some observers. The bottom line once again is that the path to truth in science is experimentation and repetition; for now, there is no experimental proof for preons, strings, and the like. The one caveat to that I will raise here is the fact that there was no experimental proof for the atomic model for thousands of years, or at least we lacked the technology to access the proof, and yet the model was eventually vindicated. "Imagination is everything" says Einstein. I would not argue that point.

The deuteron has one proton and one neutron, and because the number of protons is still the same (one), this is still the element hydrogen but is called deuterium, which is an isotope of hydrogen due to the added neutron. (Tritium is also an isotope of hydrogen, containing the single proton and two neutrons. It is vital in nuclear fusion processes, including thermonuclear weapons.) We think of the neutron as also being composed of three quarks, in this case two Down quarks each with -1/3 elementary charge, and one Up quark, with +2/3 elementary charge. Therefore, the total electrical charge of a neutron comes to zero, and we know the neutron has zero electrical charge or is termed "neutral". Quarks are confined within the neutron as well, and it is worth repeating that this aspect of gluon exchange and "color charge" has nothing to do with color or charge as we typically think of the terms. I also want to introduce the distinction between splitting protons and neutrons into free quarks, a feat that is not possible at the moment and may never be, with the process of *nuclear decay*. The latter is a very different process. For example, a free neutron outside the nucleus is actually not a very stable entity, and it decays into other particles (*not* free quarks) with a mean lifetime of about 15 minutes and a half-life of about 10 minutes. We will discuss this process in particular at length when delving into the weak nuclear force. Now we are ready to explore the higher elements, and in that adventure gain much more insight into the strong nuclear force.

The nucleus of Helium, called an alpha particle when associated with nuclear decay, has two protons and two neutrons. Immediately we see there is a problem. The helium nucleus is quite stable, and yet it has two like-charged protons which repel each other via the electric force. How is it able to stay intact? The problem gets steadily worse the deeper we go into the periodic table, as sequential elements keep adding yet another proton. Eventually we reach uranium, the last naturally occurring element, with a total of 92 protons, all of which repel one another. The answer to the riddle of nuclear stability is that there must be a force stronger than the electric force holding the nucleons together, and this force is termed the "residual strong nuclear force" since it exists *between* nucleons rather than *within* an individual proton or neutron. Though in relative terms we have not known about this force nearly as long as gravity and electromagnetism, we have discovered quite a number of its properties nonetheless.

The residual strong nuclear force is an extremely short range force, and by that we mean the strength of the force drops off far faster than the inverse square laws of gravity and the electric force, which are deemed long range forces since they extend through space over much larger distances. Since the like charged protons repel one another, the strong nuclear force (SNF) must be attractive at short range, and it turns out to be up to hundreds of times stronger than the electrical force within this range. Hideki Yukawa, in the late 1940s, predicted that this force was communicated through the exchange of pi mesons, or pions as they are usually called, and this has been verified experimentally. Also from experimental results, as well as theoretical predictions, the range of the SNF has been found to be about 3 Fermis, which means if nucleons are separated by more than that amount of space, the SNF is essentially zero and the electrical force takes over. Interestingly, there is a minimum range of proximity as well, located just over a half a Fermi, wherein the SNF becomes a force of *repulsion* between nucleons. This prevents the nucleons from completely squashing one another and thus gives the nucleus its size. One can begin to appreciate the tight parameters within which the residual SNF operates!

Actually, this idea of the nucleons not getting too crowded is yet another extension of the Pauli Exclusion Principle, set forth by Wolfgang Pauli in the first half of the 20th century, and which was applied to electron structure within atoms. The Exclusion Principle states that no two electrons can occupy the same quantum state. This is a "rule" which has profound implications for the structure of matter: since we cannot jam electrons all together into one orbital, the result is several different orbital shells with distinct shapes, sizes, and bonding characteristics. Digging deeper, since it has been proven that atoms are 99% empty space, the only reason you are able to sit on a "solid" floor is because of the Exclusion Principle – you are really sitting on an electron cloud repulsive force field made up of the atoms in the floor. Each element, substance, compound, and so forth has a unique electronic structure, so if we take water as an example, apparently its outer electron cloud repulsive forces are not enough to sustain the weight of a human being. Thus, we cannot walk on water, even if we account for the surface tension due to intermolecular bonding that is so important in capillary action. The notion of quantum states has a precise meaning, but for now it is sufficient to note that nucleons obey the Exclusion Principle just as the electrons surrounding the nucleus do.

Within the range of roughly .5 to 3.0 Fermis then, we have found the residual SNF to be attractive, stronger than the electrical force, and responsible for holding nuclei with multiple protons and neutrons together. Up to about the element calcium, the number of protons is usually balanced by an equal number of neutrons (with the exception of isotopes). However, once we get past elements with 20 protons or so in the nucleus, things begin to unravel a bit, and we can see we have a problem. It is not difficult to see why this would be the case: the nucleus is getting bigger due to having so many protons and neutrons, and since the residual SNF is extremely short range, the battle between that force and the electrical "Coulomb" repulsion of the protons is getting to be a bit lopsided in favor of the electrical force unless we bring in extra reinforcements to strengthen the SNF. This is exactly what nature has decreed, and the reinforcements come in the form of "extra" neutrons to bolster the nuclear force side of the equation. By the time we reach uranium, in its most natural state, called U-238, the uranium nucleus has the required 92 protons (to be uranium) and *146 neutrons*.

The number 238 is arrived at of course because it represents the atomic mass number, which is the sum of the nucleons in the atom (electrons are roughly 1/2000 the mass of a proton or neutron, so their mass is not a significant contribution to the mass of the atom). Even with all the extra neutrons, the repulsive effects of 92 protons in close proximity makes the uranium nucleus highly unstable, which is why it (and most other elements with high atomic numbers) decays until it reaches a more stable state. As mentioned previously, we will look at radioactive decay in much more detail when we tackle the weak nuclear force, but for now perhaps you can see why some elements are more unstable than others. Incidentally, the term "radioactive" comes from the Curies, Marie and Pierre, and traces its roots back to Latin like most every word in English it seems. The point I want to emphasize is that it has nothing to do with radios! Since Madame Curie has entered the stage again, perhaps now is the time to grant her the ovation she deserves, because she is the only scientist to hold Nobel Prizes in both chemistry and physics. One other incidental, before we investigate the final aspect of the SNF that has such monumental consequences for the human species. Gravity cannot be the attractive force which holds nucleons together, since it is simply far too weak and the masses involved are many orders of magnitude too small to account for nuclear stability. If we were to scale things in a rough order of magnitude manner, the SNF would be 100 times stronger than the electrical force, which in turn would be about 10^{40} times stronger than gravity! Onward to that last aspect of the SNF up for discussion: fission and fusion.

With the ongoing synergy between advancing science and new technology, the 1930s brought in a vast increase of experimental methods designed to further explore the atom, and in most cases that meant the nuclei of atoms. In a vein similar in approach to Rutherford's alpha particle scattering technique (that you may recall was used on thin gold foil to deduce the positively charged, dense nucleus of atoms), people began firing subatomic projectiles at the nuclei of various atomic elements in an effort to "see" what was going on inside. With Chadwick's discovery of the neutron, Enrico Fermi (among others) came up with the idea of using neutrons as the "bullets" to probe the nucleus. This strategy had the distinct advantage of avoiding any electrical "Coulomb" interaction with the positive nucleus due to the neutral or zero charge of neutrons. The preferred target was uranium, since it has 92 protons in its nucleus and thus the titanic battle between the SNF and the electrical repulsion amongst the positive charged protons was nearly an even match, rendering the uranium nucleus unstable.

It was toward the end of this decade, 1938 into 1939, that a group of scientists - Meitner, Frisch, Hahn, and Strassman – made a momentous discovery. Detailed analyses of the products resulting from bombarding the uranium nucleus with neutrons revealed that what had actually happened was the splitting apart of the uranium nucleus into lighter elements, such as barium and krypton. In addition to these lighter elements, three free neutrons were released per fission event. This discovery on the face of it may not seem that important, but it initiated a series of events that radically changed the outcome of World War II and all the decades which followed. It is a matter of some historical debate, and interest, how it was that such a discovery somehow never quite found its way to an actual application under Hitler's fascist regime. Some may claim that it was Hitler's own propensity for absolute secrecy that is anathema to scientific progress; others would argue it was by sheer good fortune and serendipity that things played out the way they did.

Regardless, the science behind the fission process is now very well known, available in thousands of libraries and textbooks, and remains a classic example of how advances in science and technology can neither be "contained" nor controlled by the scientists and engineers who make those discoveries. In the case of nuclear fission, it wasn't for a lack of trying.

To understand why this discovery has such huge ramifications for the human species and the planet, we need to delve into the mass-energy structure present in the nuclei of all the various elements in the periodic table. For the uranium nucleus, the binding energy per nucleon is relatively small, since the electrical repulsion among 92 protons is so substantial, causing the nucleus to be quite large in extent, which in turn causes the residual SNF to markedly decrease in strength because it is such a short range force or interaction. When we speak of binding energy per nucleon, we are using this as a measure for how much energy is needed to break the SNF bonds between nucleons apart, allowing the electrical force to now dominate and potentially liberate the nucleon(s) from the nucleus. It becomes evident that the uranium nucleus is a prime candidate for an "intruder" to bust it apart, since it is already highly unstable. *The graph of binding energy per nucleon is thus the key to the entire fission (and fusion) puzzle, because it is not a horizontal line across all the elements.*

Sometimes using extreme cases when trying to unravel a problem is a powerful tool. That is the methodology I will employ here to illustrate why the aforementioned graph could never be a horizontal line across all elements. One might assume that by starting on the extreme far end of the natural elements, namely uranium, the binding energy per nucleon would gradually increase in a diagonal (negative slope) line until it maximizes at hydrogen at the other end. But wait! We are forgetting that for the lighter elements with far fewer protons and neutrons, the nuclei tend to be smaller in extent, hence the SNF interaction is greater in strength because the nucleons are at shorter range with one another, while the electrical force of repulsion is decreasing since there are fewer protons in these nuclei. This would suggest there is a point where the two forces, SNF and the electrical force, might balance one another, and this would be the most stable nucleus with the highest binding energy per nucleon because it is so reluctant to break this equilibrium. Correct, and that point is an isotope of nickel, Ni-62. You may have read somewhere that iron is the most stable nucleus amongst the elements, and when we look at this process from a mass per nucleon perspective, that is in fact the case. When we assemble this puzzle in its entirety, we will then see why the central or inner core of planets, including our beloved Earth, is typically made from iron and nickel (and perhaps cobalt along with trace amounts of other elements). The wish here is to be accurate, but not at the expense of obscuring what we are trying to figure out. Let's not get lost in the weeds, at least for the moment!

So, what does the binding energy per nucleon graph look like? From our discussion above, see if you can figure it out – draw a rough sketch of it with binding energy per nucleon on the Y-axis and element number on the X-axis. Then look it up online and see how close you came to the proper (experimentally verified) graph. Too often, both in conversation as well as teaching, answers are given far too soon, and wonderful questions, tangential and otherwise, are squelched and the discussion stops. I will assume you are looking at that graph now, and like every good graph in physics, it tells us a great deal about how nature operates. Recall that when we fire those neutral projectiles called neutrons at the uranium nucleus, sometimes the end product was discovered to be lighter elements, such as barium and krypton. Looking at the binding energy per nucleon graph, notice what happens when we move toward the left from uranium toward these two elements: the binding energy per nucleon gets bigger! Apparently, when we fission uranium, we are releasing energy that was previously stored in its nucleus. And when we calculate how much energy is released per fission event, things start to get really interesting, because *compared to the chemical energy (per molecule) when exploding dynamite, this nuclear fission process per event releases millions of times more energy.* One more detail emerges from this fission process, an item Enrico Fermi (and others) pointed out, and that eventually led Leo Szilard and Edward Teller to visit Einstein and urge him to write a letter to President Roosevelt, asking the government to dedicate substantial money and resources to investigate this new discovery.

Review the fission process and I am betting you will spot this salient detail immediately: each fission event produces a small number of free neutrons, as well as the byproduct of lighter elements. Hence, if we have three free neutrons per fission, and we assume each free neutron can produce another fission event, we can go from one event to three, to nine, to twenty-seven, to eighty-one, and so forth rapidly. If such a chain reaction can be accomplished, then huge amounts of energy would be released, all at once. Einstein agreed to sign the letter to Roosevelt, and due to his worldwide fame at this time, his request of the President carried immense weight. Thus, the Manhattan Project was born, and with it the nuclear age began with an all-out effort to secretly build the first fission bomb, with the hope of bringing the war to an end much sooner than anticipated.

I am going to pause here for a moment, and leave the field of physics to discuss a matter of historical significance that reaches across many disciplines and speaks to our survival as a species. What is the responsibility of the scientist when doing research, or any type of work, that has the potential to do great harm while also possibly being of tremendous benefit to humankind? As I referred to in the Foreword to the book, I taught a course once, along with a colleague, that went by the title "Science, Technology, and Society". Within months it was oversubscribed because students had a passion for discussing this issue of responsibility in the sciences. We debated many issues, whether it be in the field of genetics, chemical waste, renewable energy sources, healthcare, nuclear power, and of course nuclear weapons. When we arrived at the issues raised by the Manhattan Project and the use of the fission bomb on Hiroshima and Nagasaki, I divided the students up into five groups. Each group chose which perspective they wanted to defend from the five options given: threaten to use the bomb but do not drop it, use the bomb on a military base, demonstrate the use of the bomb on a deserted island, keep the bomb secret at all costs, or use the bomb on industrial sites with high civilian populations. Each participant was asked to approach this topic with the mindset that the event has not yet been decided. The driving questions was: "What would you have done had you been in the shoes of the scientists and engineers who built the bomb?" When listening to the arguments raised from each perspective, I was constantly reminded of the merits in each of the perspectives, and the passion, complexity, and difficulty in making such a consequential decision.

Students were so articulate, and had done such remarkable research on the topic, that I found myself questioning every perspective – particularly the one I had held for a long time. Einstein, in later life, is reported to have said that he never would have signed that letter had he known that it would result in the bomb being used on Japan, rather than against fascist Germany, which at the time seemed a far greater threat to develop the bomb first and have it used against the allies.

I wrote a poem a long time ago addressing some of this, and I include it here because how often does one see a poem in a physics book? Feynman once said or wrote, I cannot remember which, that "poets do not write to be understood". Besides having a good laugh at that impudent remark, I also saw some truth in it, because I confess much of abstract art and poetry leaves me completely unaffected. A thick red line across a canvas does not speak to me; neither does a pile of words add up to any impact if they are all shrouded in some deeper meaning known only to the author (and sometimes not even to the author). Give me Picasso's *Guernica* – there is abstract art of incredible power, or Maya Angelou's *"I Know Why the Caged Bird Sings"*, or the second movement of Beethoven's *Piano Concerto No. 5*, known as the *Emperor*. All of these and many more, particularly music in all its forms, are lasting gifts of the human spirit. This little poetic interlude will therefore have a meaning that is plain to see, and though galaxies apart from the masterpieces alluded to above, gets to the point of the matter. Incidentally, an MeV is a unit of energy often used in atomic and nuclear physics because it is easier to use than the Joule: 1 eV = 1 electron volt = 1.602×10^{-19} Joules. We don't like such small exponents in common usage, so we convert to electron-volts. The M is a metric prefix that stands for "mega" or million. Thus, MeVs is pronounced one letter at a time: M-e-V(s).

> Who knows the secret of hydrogen?
> Who thinks he knows and sees?
> With the binding of matter to matter,
> How many thousand MeVs?
>
> Who goes to split the proton,
> And finds three quarks inside?
> People of science, the horse is there
> But the state will take the ride.
>
> Einstein traces the contours
> And predicts the selling of science.
> Come all you fools behind locked doors –
> Come shout your angry alliance.
>
> For freedom is granted until the find,
> But then it leaves your hands.
> Transformed it travels all over mankind;
> God save us all when it lands.

One can reasonably argue that even in the madness of a nuclear arms race which, if unleashed, would end all human existence now and forever, there is also the potential that such a capability elicits enough terror to actually prevent a third, catastrophic world war. Nuclear fission also has some benefits, in that its development had a large impact on advancing medical technology that ultimately saves lives. The harnessing of fission power, though controversial with good reason, still has provided electricity to many millions of people. I'll conclude this interlude with the simple plea for all scientists, engineers, teachers...indeed *all* professions: please pay careful attention to the fruits of your labor, and at the very least think through *all* the ramifications.

Returning briefly to the Manhattan Project, there are dozens of outstanding books on the subject, describing how General Groves oversaw thousands of scientists and engineers at Los Alamos under the direction of lead scientist Robert Oppenheimer, while at the same time Fermi and others labored underneath the football stadium at the University of Chicago in an effort to achieve the first sustained fission chain reaction. There is no need here to drill way down into the details of this "best kept secret" enterprise, but there are some points worth emphasizing. First, the uranium isotope U-238 is naturally occurring, although quite rare, but it is not likely for a number of reasons to undergo fission when neutrons are fired at its nucleus. The isotope U-235, on the other hand, has a far greater propensity to fission, and it is a high percentage of this isotope that is used in fission weapons. Nuclear power plants, on the other hand, use the isotope at a far smaller concentration, to the point where the rapid chain reaction necessary for a fission bomb is not possible. I hasten to add, however, that a runaway chain reaction in power plants can still occur over a period of days, known as a "nuclear meltdown", and carries with it disastrous consequences, such as were (and still are) observed in the Chernobyl accident.

The three (or so) free neutrons that are produced in each fission of U-235 have three possible fates: escape the sample altogether, get absorbed by a nucleus, or cause another fission. The essence of the fission bomb design was to engineer the process so that the last option (another fission) becomes much more likely than the other two. Whether it was the gun-type U-235 bomb or the implosion type plutonium weapon, the uncontrolled chain reaction was the key to unlocking the immense energy inside the nucleus. To accomplish this, enough U-235 has to be recovered from the natural U-238 found in rocks, a process usually involving centrifuging the element (diffusion can also be employed) so that the heavier isotope is separated from the lighter one. Plutonium can be obtained from uranium via a nuclear reactor in a manmade process. This is why when you read about verifying possible nuclear weapon violations, centrifuging operations are a key matter of interest, as well as how nuclear power plants are processing their uranium.

The refining process for the Manhattan Project was done at Oak Ridge in Tennessee and also at the Hanford plant in the state of Washington. I recall reading somewhere that when Fermi finally was able to sustain the first chain reaction underneath that football stadium, he telephoned Los Alamos and said "the Italian navigator has landed in the free world", code for "we have achieved a sustained chain reaction". This feat was accomplished not without some real trepidation, because even with the use of control rods to slow down the chain reaction, nobody had done anything like it before, and it wasn't clear what might happen once the reaction started to occur. While all of this has long since been declassified, around 1940 it was extremely dangerous information indeed. Now one can find this type of weapon design in many public places, for better or worse. I should add, however, that it took thousands of the best minds in science and engineering several years to design the first fission bomb, so the odds of a wandering maniac doing this in his basement are very long indeed. Still, nuclear proliferation is the proverbial genie out of the bottle, and it remains an extremely complex and difficult issue.

One can come at this fission process from the angle of mass rather than energy, but the end result is the same. For example, the mass per nucleon graph would look like the binding energy graph turned upside down. From that perspective, when uranium undergoes fission, the total mass content of the products is less than the original mass of the uranium nucleus, hence the phrase "mass defect" in the fission process. This is a striking verification of Einstein's most famous (and most misunderstood) equation: $E = mc^2$. The missing mass shows up as a tremendous amount of energy, since c = the speed of light = 3×10^8 meters per second. One can see that the energy content in just one kilogram of mass is enormous: 9×10^{16} Joules! This is the juncture where we need to dispel a few misnomers and myths.

First, mass is not "converted" into energy as is often written and said. The meaning of that famous equation is somewhat different: it is telling us that mass *is* energy, and a lot of it. This is akin to when reports of "zero gravity" are made as astronauts whiz around the Earth in orbiting capsules or space stations. That is also demonstrably false, since if there truly was zero gravity at this location, the spacecraft or space station would not be in orbit around the Earth, it would fly off tangent to its circular path and be forever lost. Of course there is gravity out there in orbit, albeit weaker than what it was on the surface of the Earth. Newton would say we are farther from the center of the Earth, so once beyond the surface the acceleration of gravity drops off as a square of that (increased) distance, but the value of "g" is still not zero. Einstein would say where there is less curvature of spacetime, as in farther away from the mass creating that curvature, there is less "gravity". Either way, what is occurring is freefall: everything is falling *because of gravity*, and at the same rate regardless of mass, so the appearance is one of weightlessness since there is no relative point of contact with anything if it is falling around the Earth as fast as you are! The final myth to address here is when I hear or read that the Sun is "burning hydrogen to form helium". Absolutely not. We figured out a long time ago that if chemical combustion was the source of the Sun's energy, it would have burned out millions of years ago. In fact, the source of the Sun's energy, and the source of energy for all stars, is not a chemical process at all – it is nuclear fusion. This was not explained fully until Hans Bethe parsed out all the details of the fusion process in the late 1940s. We have reached a perfect segue for our transition to the third fundamental interaction in nature, the weak nuclear force, because it is the weak interaction that allows stars to shine and is so instrumental in the nuclear fusion process. Once we fully explore the weak force, the process of nuclear fusion will be explicated in the process. Onward.

WEAK NUCLEAR FORCE (INTERACTION)

It seems the weak interaction gets the least notoriety of the four fundamental forces, perhaps due to the label "weak", or possibly because it isn't as obvious or as observable as the other three forces (though one could argue that the SNF is not readily observable as well). That is a bit unfortunate, since without the weak interaction, the Sun and all stars would not shine, and we would therefore not exist. Obviously then, the weak interaction plays a vital part in the process of nuclear fusion, which is why we left the fusion process to this section of the book. To fully appreciate the crucial role this interaction has in our universe, we need to first do some "trail clearing" to open up the path to this particular vista. This involves a deeper understanding of nuclear decay, and for our purposes entails a focus on the three main types of decay schemes, all of which serve as nature's attempt to stabilize the nuclei of those elements which are teetering on the edge of instability. Those three modes of nuclear decay are alpha, beta, and gamma radiation. As we shall soon see, one of these modes is governed by the weak nuclear interaction.

When we look at nuclear decay, we see that these transformations within the "parent" nucleus either eventually result in more stable "daughter" nuclei of different elements, or in a lower, more stable energy level within the nucleus of the same element. One term you may be familiar with is the concept of half-life, which is the amount of time it takes for exactly half of a radioactive substance to decay into a specified product. Alternately, we can use the idea of the amount of activity as the measure, and define half-life as the amount of time for the radioactive activity to drop to half its original value. As an example, the half-life of carbon-14 (C-14 = 6 protons and 8 neutrons) is approximately 5700 years, meaning half the amount of C-14 in a specimen will turn into nitrogen (the product) every 5700 years. Exactly how that occurs we will discuss shortly, although I want to emphasize here that this is a statistical process based on probability, so while we can measure the amount of the substance, we have no way of predicting or knowing *which* carbon atoms will decay at any given moment.

We know that a small amount of C-14 is present in all living carbon-based life forms, including humans of course, *in exactly the same percentages*. Through photosynthesis, plants absorb carbon dioxide and also small amounts of C-14 present in the atmosphere, which animals ingest, which is why we have it in our bodies as long as we live. The amount of C-14 remains constant while we are alive. Once we die, however, we no longer take in the C-14, and since it is continually decaying, the percentage amount now starts to decrease. Suppose we investigate the age of a fossil and find that it has about half the amount of C-14 now as it did when it was living. That means the organism died about 5700 years ago. If it has 25% of what it had while living, then it has gone through two half-lives, which means it died about $2 \times 5700 = 11,400$ years ago. This is how we use carbon dating to determine the age of fossils, and it works very well up to about 10 half-lives, or 57,000 years or so. After 10 half-lives, there is too little C-14 left to measure with a high degree of precision and reliability, and other (much longer half-life) elements need to be used. Again, we will return to C-14 shortly, because it is so important to life and it is intimately involved with the weak nuclear interaction.

In alpha decay, the nucleus moves toward more stability by ejecting a helium nucleus: two protons and two neutrons. Obviously, alpha decay cannot occur for hydrogen (only has one proton) or helium (the nucleus is already an alpha particle by itself – nothing would be left!). In fact, alpha decay is far more probable and prevalent in elements of higher atomic number than iron or nickel. This is related to the binding energy per nucleon discussion, but it is most easily understood by recalling that when large numbers of protons reside in a nucleus, the electrical repulsive force is battling the short-range residual SNF, and it is nearly winning the battle because as the nucleus gets bigger, the electrical force of repulsion between like charges doesn't change much but the strength of the attractive SNF drops off precipitously. Therefore, in a manner similar to the fission process, the decay goes in the direction of more stability. Let us look at one example of alpha decay that has related significant health issues: the U-238 nucleus. This uranium isotope occurs naturally in samples of granite and other substances, and its decay accounts for almost half the heat generated in the Earth's interior!

The first alpha decay of U-238 results in the element thorium. Recall there are 92 protons and 146 neutrons in U-238, so in its goal of trying to achieve more stability by ejecting an alpha particle (2 protons and 2 neutrons), a new element is obtained, with 90 protons and 144 neutrons: Th-234 (Thorium). This particular portion of the decay has a half-life of over four billion years, which is the approximate age of the planet itself. That is why there is still uranium around today! There are a couple of intermediate steps (beta decays – discussed in detail soon) which, within a matter of days, bring us back to a slightly different isotope of uranium, U-234. An alpha decay then occurs again, leaving us with Th-230 (2 less protons means we are at Thorium, 4 less nucleons means we have an atomic mass of $234 - 4 = 230$). This decay half-life takes just under a quarter of a million years. Another alpha decay results in radium: Ra-226, a process with a half-life of about 77,000 years. The next alpha decay is the one that has been in the news the past few decades or so, as radium decays into radon gas: Rn-222. Radon alpha decays to polonium with a relatively short half-life just under four days, and polonium undergoes alpha decay to end up at lead, which undergoes a series of processes and usually ends up as the isotopes of lead Pb-

206 or Pb-208. That is where the decay scheme of uranium reaches its terminus, and the point at which there is much more stability within the nucleus.

At this juncture it is important to discuss the health ramifications of alpha emitters, in particular radon gas. Since many foundations built in certain regions of the globe involve digging through rock formations that contained small amounts of U-238 billions of years ago, at this point in time the possibility of radon gas being present and trapped within these foundations cannot be ignored. Alpha particles, on the atomic scale, have a large mass (4 nucleons) and a substantial charge (+2 elementary charges). This means they will interact with matter extremely rapidly, and in one sense that is a good thing: alpha particles do not penetrate barriers such as air, paper, skin, or most anything else before being completely absorbed. So, what makes alpha emitters a possible danger to humans? Two factors: radon is a gas that has no odor, and the fact that we breathe. If enough alpha particles are ingested into our lungs, they possess enough energy to disrupt DNA sites and cause lung cancer. This is why we have developed ventilation systems to rid any enclosed areas of residual radon gas, as once vented into the atmosphere it dilutes and gets absorbed rapidly in a manner not dangerous to humans. The key is to keep alpha emitters outside the body; our skin alone will protect us, and we are in no danger unless we breathe it in as a gas or ingest it by eating or drinking from a contaminated source. That being said, alpha sources have several practical applications that accrue much benefit. They are used in smoke detectors, spaceflight, pacemakers, and ironically enough, to treat cancer. When small amounts of an alpha source are *very carefully* targeted into cancer cells, it kills them without having the energy to disrupt nearby healthy cells.

Why did nature choose alpha particles to be emitted from unstable nuclei? Why not single protons or some other combination of nucleons? The basic answer to this riddle is that helium nuclei are the preferred candidate in alpha decay because it is the "easiest" path to take, meaning it takes lesser amount of energy through this type of emission than any other. And yet the process of alpha emission from nuclei is absolutely fascinating, because by the laws of classical, Newtonian physics, it should *never* happen. Suppose, like Sisyphus, we are trying to push a big rock up a hill. In keeping with the futility personified in the myth, let's suppose we have 100 units of energy to give to the big rock, but the height of the hill requires 200 units of energy to reach the summit and have it roll down the other side. It appears the situation is hopeless, because no matter how hard we push, there isn't enough energy available to get the rock over the "potential hill". From a macroscopic point of view, the hopelessness of Sisyphus is well earned: it cannot be done. But we are assuming that the same set of rules and laws that work for "ordinary" sized objects also apply to the microscopic, or atomic, world. Not so! It may not be any consolation to Sisyphus, but there is a way, at the subatomic level, to have that "big rock" tunnel through the barrier and escape out the other side!

Quantum tunneling is one of the more bizarre consequences that flows from the equations of quantum physics, and it is exactly how alpha particles escape the nucleus of atoms such as uranium. A little calculation shows that alpha particles bouncing around inside the nucleus of heavy elements such as uranium do not possess enough energy (classically) to escape the nuclear potential well they are trapped in, and thus appear to be confined within that nucleus forever. In the strange and wonderful quantum world, we employ everyday analogies as often as possible in an attempt to illuminate what few if any people truly comprehend. Instead of thinking about things like electrons, protons, and alpha particles as tiny spheres (miniature golf balls?) that are localized at a specific point in space, the Schrodinger equation in quantum mechanics assigns a "wave function" to them, whereby they spread out like a wave, with the peaks of the wave being places where there is a high probability of finding the particle, and the troughs a low probability. This is more a mathematical formulation than a physical wave; it is a methodology that involves *probability*, and it is undergirded by the Heisenberg Uncertainty Principle (HUP). One version of the HUP states that there is no experimental method which will identify both the position and momentum (mass times velocity) of any particle, *simultaneously*, with unlimited accuracy. As soon as you try to locate the position of an electron or any subatomic particle by scattering a photon of light off it, you invariably alter its momentum in doing so. This isn't some failure due to the practice of poor experimental method, it is a principle inherent in nature! *The act of observation changes that which we are observing.* At the macroscopic level, these changes may be undetectable, but inside the atom, all bets are off.

Now we return to alpha decay, and apply the principles of quantum physics just described. Be advised, though this may seem like Alice in Wonderland territory, if quantum physics did not work, neither would this computer I am typing on, along with millions of other applications which flow from its predictions. The wave function of the alpha particle trapped inside the nuclear potential well has a tiny probability of existing beyond the "walls" of the nucleus under the proper conditions – it can tunnel through the wall if the wall is not too thick! Even though the probability of a single tunneling event for a given alpha particle is typically very small, there are billions of collisions with the walls every second, so the cumulative probability allows occasional escapes to occur, even though we cannot predict which particle will be the escapee. The weak nuclear interaction is not involved with this type of decay, as at its root alpha decay is fundamentally a war between the electromagnetic force and the residual SNF. This is how alpha decay occurs in the wild world of the atom.

The second type of nuclear decay we will explore is gamma radiation, and in some ways it is the easiest to visualize and explain. In gamma decay, the nucleus of the atom changes energy state to a more stable one, much like how electrons jumping between orbital levels in the atom emit or absorb photons with an exact energy that equals the difference in energy between the orbital levels involved in the transition. Therefore, in gamma decay, the number of protons and neutrons within the nucleus remains exactly the same, and the element does not change into a different element. The result of gamma decay is a more stable nucleus at a lower energy level than before, and the "missing" energy is carried off as a gamma ray *photon*. Gamma rays are extremely high energy photons – the highest level illustrated in the EM spectrum we looked at many pages ago. As they have no rest mass and no electric charge, gamma ray photons penetrate through materials as dense as lead, inches thick. Starting with high frequency, high energy ultraviolet EM, on up through X-rays and then gamma rays, the increasing penetration power of these photons has major implications for human health.

The third type of nuclear decay, and the one mediated by the weak interaction, is beta decay. In a nutshell, beta decay is nature's clever scheme which allows a proton to change into a neutron, or a neutron to change into a proton. To understand this process, we return to the internal structure of the proton, 2 Up quarks and 1 Down quark, and the neutron, 2 Down quarks and 1 Up quark. Therefore, for a proton to turn into a neutron, one of its Up quarks has to change into a Down quark, and for a neutron to change into a proton, one of its Down quarks must change into an Up quark. Looking at the latter process first in which a neutron turns into a proton, this is called beta-minus decay, with the neutron turning into a proton, a high speed electron called a beta particle that is ejected from the nucleus, and a third ghostly particle called an antineutrino. It is a matter of historical and even philosophical interest that the neutrino, roughly translated as the "little neutral one" in Italian, was predicted to exist many years before we were actually able to detect it experimentally. The basis for that prediction will be discussed in the final section of the book.

Returning to our C-14 discussion, we now can analyze its decay because it occurs by beta-minus decay. Note that a proton is gained here, so the element does change to the next higher element, in this case, nitrogen. The total atomic mass stays about the same, since exchanging a neutron for a proton has very little effect on the mass of the atom. When we write nuclear decay processes, we have developed a shorthand for it as follows. Here is an example of alpha decay: $_{92}U^{238}$ ➔ $_{90}Th^{234} + {}_2He^4$ ... where the last bit is the helium nucleus or alpha particle. The example for the beta-minus decay of C-14 looks like this: $_6C^{14}$ ➔ $_7N^{14} + {}_{-1}\beta^0$ + antineutrino ... where the ß(beta) particle is the high speed electron. There is a symbol for the antineutrino as well, often written so that it looks like a small case script "v" with a line over it to indicate it is an antimatter particle. These are just shorthand ways to write exactly the processes we have been discussing. Incidentally, the gamma decay scheme would keep both sides of the decay equation the same, with the addition of a gamma ray symbol of zero mass and charge on the right side representing the high energy photon coming out of the nucleus.

Therefore, in beta-minus decay, nature has figured out a way to balance the number of protons and neutrons by transforming carbon-14 into nitrogen. Now let's turn our attention to the beta-plus process, turning a proton into a neutron. Predictably, it looks like the beta-minus process, just in reverse. Here, a proton changes into a neutron by transforming one of its Up quarks into a Down quark, and the result is a neutron plus a positron (positive electron – the antimatter counterpart to the electron) plus a neutrino. It is exactly this decay scheme which becomes so critical in the nuclear fusion process that powers the Sun and all stars, and therefore gives us the possibility of life on Earth! We will look at nuclear fusion first, as promised a while back, and then tie all the strings together to see how this tapestry makes sense.

Harkening back to the binding energy per nucleon graph, we have seen how the fission process releases energy when we start from the uranium end of the periodic table and work our way back to iron and nickel. Perhaps you have already seen that we could just as easily start at the hydrogen end of the periodic table and work our way forward to the higher elements, up to iron and nickel, and in this case energy will be released as well – in fact, the fusion process releases more energy than fission on average, per event. Accomplishing this task is no easy matter, however, because to join two nuclei together, to "fuse" them into a higher element, requires that we overcome the massive electrical repulsion force between the positive charged protons. This in turn requires huge temperatures and densities in order to slam the nuclei together hard enough – get them close enough to each other that is – so that the residual SNF can take over from the electrical force. We have seen how temperature is directly related to the energy of motion, kinetic energy, of a substance. The greater the temperature, the faster the molecules move on average.

On Earth, one way to cause this to happen is realized in the design of thermonuclear weapons, in other words, hydrogen bombs. In those weapons, a small fission bomb is used to generate the enormous temperatures necessary for the much larger, fusion reaction to take place. This is why hydrogen bombs followed closely on the heels of the fission bomb, coming to fruition in the early 1950s and tested with often tragic consequences until the Test Ban Treaty was signed in the 1960s. The destructive power of these weapons is almost impossible to exaggerate: the largest one to date has the equivalent energy of 50 million tons of TNT, and if detonated on the eastern seaboard halfway between Boston and Washington DC, it would wipe both of them out and everything in between. Mutual assured destruction is MAD indeed, and the argument rages on as to whether such doomsday weaponry saves us from world wars or brings us closer to extinction. Yet like so much of technology, there is a promising positive side in trying to control the process of nuclear fusion here on Earth. The way we try to cause nuclear fusion of hydrogen atoms in a controlled manner involves irradiating deuterium pellets with synchronized lasers, or containing the hot plasma in a magnetic field (since at such high temperatures any known material would vaporize). Both engineering processes face formidable challenges, and to date the amount of energy released in the fusion process is less than the amount of energy required to stimulate the process in the first place. In brief, it is not yet practical. But what a transformation of the human species and our trajectory for survival this would be if this enterprise succeeds! The fuel for the fusion process could be water, or anything containing hydrogen. All our energy needs could be satisfied with no radioactive waste products associated in the process. All waste products could be recycled, eliminating the need for landfills and eradicating billions of tons of microscopic plastics invading our lungs and oceans. Climate change would be drastically altered, for the better, with the virtual elimination of our dependence on fossil fuels and the carbon dioxide they produce. The stakes are so high that this research simply must continue. The choice is in the hands of leaders, many of whom are elected by us. *Therefore, we must do our homework, and vote accordingly.* Now let us turn our attention to the stars, and pull all the pieces of the puzzle together.

We know from spectral analysis that stars like our Sun are principally composed of hydrogen in a plasma form, where the temperatures are so hot that electrons have been easily stripped away from the hydrogen nuclei. The process of turning hydrogen into helium releases fantastic amounts of energy vis a vis the binding energy per nucleon graph and the equivalence of mass and energy. This outward energy flow creates a pressure that balances the tendency of mass to collapse in on itself due to the extreme curvature of spacetime, or gravity if you will, and thus stars like our Sun are called main-sequence stars that will continue this process for billions of years on average. Eventually the hydrogen runs out, and the star may begin to fuse helium into the still higher elements like carbon, oxygen, nitrogen, and so forth, up to iron and nickel. Are you beginning to see why the core of most planets and "burnt out" stars are composed of those metals? When the fusion process fizzles out, due to lack of an available fuel that will *release* energy in the process, the collapse of the star commences. What happens next depends on several factors, but chief among those is the mass of the star.

The star could restart the nuclear fusion process upon collapse, because the collapse might generate high enough temperatures, densities, and pressures to do so. It might turn into a "red giant", like we think our Sun will eventually do billions of years from now, and swell out to a size that gobbles up its nearest planets (and we will likely be one of those planets). One would hope by then we will have populated the galaxy in more hospitable places! Other scenarios include shriveling up into a white dwarf, or collapsing into a star having only neutrons packed in to astounding densities, and spinning around at phenomenal rates, sending out pulses of energy all the while (pulsars). And of course if the mass of the star is big enough, the collapse might never end and we have a massive black hole, with an event horizon beyond which all physical laws we know no longer seem to hold. All of these scenarios are occurring all the time, just as those ghostly neutrinos are passing through the Earth (and us) by the billions every day, even though we cannot see them. But there is a central question you may have thought about while reading through the last page or so, with a resolution that is tied to the weak nuclear interaction.

If the Sun fuses hydrogen in plasma form, consisting of protons only, into helium, consisting of two protons and two neutrons, *where do the neutrons come from*?!

The resolution, of course, is that the neutrons arise from the beta-plus process, in which a proton transforms into a neutron. It is this intermediary process which produces isotopes of hydrogen, principally deuterium, which in turn through a complex series of events will fuse into helium. The whole deal is governed by the weak interaction nuclear force, the one that takes one of the Up quarks in the proton and changes it into a Down quark, making it into a neutron. This process was thought to be governed by the exchange of massive particles, much more massive than the proton or neutron, with the less than exotic labels of W^+, W^-, and Z particles. The work of Glashow, Salam, and Weinberg in this field earned them the Nobel Prize in 1979, and in the process of their efforts, the weak interaction and the electromagnetic interaction were unified as manifestations of one force, now dubbed the electroweak force. The W and Z bosons were seen to share many characteristics with the photon (other than mass of course), and as we shall soon see, the photon is the key player in the electromagnetic force. Lest we forget that experiment provides the final answer, both the W and Z particles were discovered at the CERN facility in the early 1980s. Once again, theorized particles found their way to experimental verification. The last remaining interaction for us to investigate we have actually already done a significant amount of traveling with – the electromagnetic force.

ELECTROMAGNETIC FORCE

We covered a lot of terrain in the section on electric charge, so there will be some overlap as we discuss the electromagnetic force in more detail. The pervasiveness of this fundamental interaction relies on the fact that all of the "stuff" that makes up matter (and antimatter) can be traced back to atoms, and atoms are held together by the electrical attraction between positive nuclei (the contribution of protons) and negative electrons. Therefore, when we study the tension in a rope, it is the electromagnetic force at the root of the strength of the bonding within the rope. The same holds true for the force of friction between surfaces, and the tension in springs: both rely on the electromagnetic force, since both are produced by the interactions among atoms. The normal, perpendicular force which holds up tables, chairs, and people is actually a manifestation of electron clouds repelling one another due to the Pauli Exclusion Principle we referred to a number of pages ago. Though the atom is almost all empty space, the electromagnetic force fields of many objects and substances gives us the sensation of sitting on something "solid". In fact, the very essence of carbon based life is held together by the electromagnetic force, in the form of the strands of the double helix in all of our DNA. Every time we blink, move a muscle, experience another heartbeat, or even think...all of that is electromagnetic in nature.

It is instructive to contrast the electromagnetic (EM) force with gravity, because in two ways the forces are quite similar, and in at least three ways they are dramatically different. Tackling the differences first, we start with the obvious fact that the EM force can repel as well as attract, whereas gravity can only attract (at least the gravity we are familiar with thus far). The second difference is quite profound, and deals with the fact that the EM force can be shielded, while we have found no substance or circumstance where gravity can be weakened in the same way. As one example of many, if we place a book between two charges, we observe that the force between the charges is markedly reduced. The same experiment involving a book between two masses has no noticeable effect on the force of gravity between the masses. It appears gravity cannot be shielded. I suppose if we reconfigure gravitation as the geometry of spacetime, then the concept of shielding should have zero impact since we live in that 4-D world. Yet when we dig deeper and view this from a quantum physics perspective, we discover that the electromagnetic force is mediated (transmitted, communicated) through the exchange of photons, and photons can be shielded or stopped by various materials. Therefore, we see the SNF as mediated by gluons within the nucleons and pions between nucleons, the weak interaction is communicated via the W and Z bosons, and now the EM interaction is experienced through the exchange of photons – light itself. We have yet to detect the gravitational exchange particle, the graviton, which is at the core of the incompatibility between the GTR and quantum physics.

The third difference between the EM force and gravity is the fact that the EM force is stronger by a factor of around 10^{40}. We have a name for 10^{100} – it's called a googol. It is sufficient to show how much stronger the EM force is than gravity by writing it in terms of sheer numbers: 10000000000000000000000000000000000000000 times stronger. I hope I counted the zeros correctly; if we are off by one or two here, I think the point is still made that gravity is much weaker than the other three fundamental interactions by many orders of magnitude (meaning, powers of ten). Charles Augustine de Coulomb came up with a quantitative formula for the force of attraction or repulsion between two stationary electric charges. This was done around the year 1800, and the experimental result he obtained is as follows : $F_{electric} = kQq/R^2$.

Here k = Coulomb's constant = $1/4\pi\varepsilon_o$ = 9 x 10^9 Newtons x meters^2/Coulombs^2; please note again how this reduces to the four basic concepts of space, time, mass, and charge. Also take note of the fact that the symbol ε_o was used by Maxwell, together with the magnetic constant counterpart μ_0, in his historic synthesis of finding the speed of EM waves as the speed of light, thus wedding the three phenomena of electricity, magnetism, and light into one fundamental process. Note how much greater the "coupling constant" k is relative to G in the formula for gravity! Coulomb's constant is 9 x 10^9 in metric units; Cavendish's G is 6.67 x 10^{-11} in metric units. Already we see the "coupling" is much greater in the EM force. The other symbols in Coulomb's Law are as follows: Q and q = the two charges, measured in Coulombs, and R is the distance between the two charges, measured center to center. As Newton demonstrated using his invention of calculus, we can show that a solid, uniform sphere of mass M can be considered to be a point particle with all the mass at its center; the same holds true for a uniform, spherical charge distribution.

Perhaps you have noticed a striking similarity between the two expressions for the force of gravity, deduced by Newton, and the electrical force found experimentally by Coulomb. *Both are inverse square laws!* One involves two masses times a constant; the other involves two charges times a constant. Both depend on distance in exactly the same way. The formulas are tantalizingly similar, and this fact probably played no small part in serving as the impetus to try to "unify" the two forces into one generalized interaction. Einstein spent decades trying to achieve this task, but despite his best efforts, he failed. In the process, he continued to resist the entire field of quantum physics, thinking that there must be some "hidden variables" we are missing that would eliminate all the uncertainties and probabilities inherent in the field. This is particularly ironic given that much of his work, such as with the photoelectric effect, lent tremendous credence to the quantum conception of the universe. Indeed, he is famously quoted as believing that "God does not play dice with the universe". The journey continues. Perhaps such a quest is futile, but the search for a Grand Unification Theory that binds all the four forces into one colossal interaction remains ongoing.

We know that stationary electric charge produces an electric field around it, with the field lines pointing outward from positive charge, in rays that are perpendicular to the surface of the charge and that spread radially outward like rays from the Sun. This superb conception of Michael Faraday is more than just a "pretty picture" – it marked a revolution in how we visualize various patterns in the natural world, and it shows (mathematically if one pursues the geometry) how the field weakens as an inverse square of the distance from the charge. We also discovered through Oersted that when this charge is set into motion, either through space or within a wire as electric current, another field is produced, namely the magnetic field. Faraday then showed that a changing magnetic field can be used to generate electricity, such as thrusting a magnet through a coil of wire or alternatively spinning a coil of wire through a magnetic field. At the close of the American Civil War James Clerk Maxwell then shows that changing electric fields can induce changing magnetic fields in such a continuous manner as to generate EM waves, and calculated the speed of these waves as the speed of light. Incredibly, the three phenomena of electricity, magnetism, and light, thought of as entirely separate for thousands of years, were now shown to be deeply connected under one unified EM force. This wasn't just a giant revolution in physics – these discoveries radically transformed the arc of human history and affected nearly every aspect of our lives.

The other "half" of the picture, magnetism, is a bit more complex to describe in terms of formulas and mathematics. As one example, the magnetic field strength due to the current in a long straight wire is proportional to the inverse of the distance a point is from the wire. Note that this is not an inverse square relationship; the magnetic field gets weaker the farther away one is from the source ($1/R$), but the field drops off slower than that of the inverse square law ($1/R^2$) that holds for the electric field produced by a stationary charge. The other feature that is critical to remember here is that while isolated charges are observed all throughout nature, isolated magnetic poles (monopoles) have never been experimentally detected. Apparently, when it comes to magnetism, nature prefers continuous loops that (by convention) start on north poles and end on south poles. I will leave it to the reader to investigate magnetic dipoles, the force between two magnets, and other such examples, in favor of a laser-like focus on one example of an electric charge moving through a uniform magnetic field.

Suppose we hold a horseshoe magnet in such a way that the north pole of the magnet is above the plane of this page, and the south pole of the magnet is below the page. Think of this page as being between the poles of the magnet, in other words. Using the convention that magnetic field lines go from north into south pole, we picture the magnetic field as lines passing down *into* the page, a situation we represent by using the letter X, since if we fired an arrow into the page, the last thing we would see is the tail end of the arrow, shaped like an X. (Field lines coming out of the page toward us are represented by dots, which are the points of the arrows coming out at us.) Now let us introduce a moving positive charge, we will use the proton as an example, and have this proton come into the plane of the page from left to right, across the page. What is the path the proton will take? How will its speed or velocity change, if at all?

It helps in such matters to slow the process down in one's mind, and take incremental steps toward the solution in order to fully appreciate all the details along the way. First, the proton is charged positive of course, so it produces an electric field around it *always*, whether it is moving or not. Second, since the proton is moving here, and we know that moving charge produces a magnetic field, we now have *two* magnetic fields: the one produced by the moving proton, and the other is the horseshoe magnet we have surrounding the page. Third, since there are now *two* magnetic fields, in essence we have two magnets in proximity to one another, so there must be a magnetic force on the proton, just as there is between any two ordinary bar magnets placed next to one another. The question then becomes: what is the magnitude and direction of this magnetic force acting on the proton?

We will establish direction first by recalling that we use a "trick" called the Right Hand Rule, or RHR, for moving positive charges or currents in a wire. To review, the RHR works as follows: hold your right hand outstretched so that your thumb is pointing perpendicular to the direction of your fingers. This should look like you are holding up your hand as though trying to stop a car or something coming at you. In this scenario, your fingers are pointed in the direction of the magnetic field lines, your thumb points in the direction of the movement of the charge (velocity "v" here), and your palm pushes in the direction of the magnetic force. Now orient your hand so that it matches the illustration we are using: the magnetic (B) field lines are *into* the page, so point your fingers into the page in a way that also has your thumb pointing toward the right side of the page. Your thumb must point to the right because that is the direction the proton is moving, in the plane of the page. Now you are in a position to figure out the direction of the magnetic force on the proton, because it is the palm of your hand pushing it in that direction. If you have done this systematically, you now know the magnetic force is pushing the proton upward in the plane of the page as it enters the magnetic field of the horseshoe.

The diagram would look like this:

```
                        XXXXXXXXXXXXXX
                        XXXXXXXXXXXXXX
Proton movement ➔ ➔ ➔   XXXXXXXXXXXXXX
                        XXXXXXXXXXXXXX
                        XXXXXXXXXXXXXX
```

Your right hand should be oriented so that your fingers are pointing down into the page (the X symbols showing magnetic field lines (**B**), and your thumb is pointing to the right for the proton movement or velocity (**v**). Thus, the magnetic force **F** on the proton is your palm pushing up as the proton enters the magnetic field of the horseshoe magnet.

Two crucial items follow from this analysis. First, notice that the velocity **v**, the magnetic field lines **B**, and the magnetic force **F** are all mutually perpendicular to each other, and all three are vectors since they obviously have direction. That is why I will put the symbols in boldface – to show that they are vectors. *Second, since the magnetic force is perpendicular to the direction of the proton, this force cannot change the proton's speed, only its direction!* One can only change the *speed* of an object if there is a net force in the same or opposite direction as the object's velocity. This is a direct consequence of Newton's Second Law: **F** = m**a**. Therefore, the proton is accelerating because it is changing its direction, not its speed. Since its direction is changing, we must continually orient our thumb accordingly to show this change in direction. Try doing this as the proton moves through the X symbols of the horseshoe magnet's field. Can you see the path the proton must take? The magnetic force on the proton will always be perpendicular to its velocity vector. I hope this might ring a bell of sorts. When a force is directed perpendicular to an object's velocity, and the speed of the object does not change, then the result is a circle. If you draw this path on the diagram in the book, it should look like a semi-circle, moving up and around counterclockwise in the plane of the page. The magnetic force is therefore a centripetal force in this example, just like we used to describe planets orbiting the Sun, or in earlier versions of the atom, electrons orbiting the nucleus.

That answers the question regarding the direction of the proton's movement, but we still need to figure out the magnitude of the magnetic force acting on the proton. We do know it is centripetal, therefore we have that piece of the formula: $F_{centripetal} = mv^2/R$, working with magnitude only here. Analysis of the factors involved would suggest that the magnetic force should depend on three factors: the amount of charge, the velocity, and the strength of the external magnetic field (horseshoe magnet here). Indeed, the magnetic force is given by: $F_{magnetic} = QvB\sin\phi$, again magnitude only here, and with the angle ø as the angle between the velocity vector and the magnetic field vector. In this case, the angle is 90 degrees since the two vectors are perpendicular, and therefore since $\sin 90° = 1$, we have the following relationship: $F_{magnetic} = F_{centripetal}$ ➔ $QvB = mv^2/R$ ... which is an incredibly useful tool in many situations, such as with mass spectrometers, because it can identify which particle we are looking at. Let's take a (return) look at two of those applications, but now we are equipped with better tools!

The Van Allen Radiation Belts are a perfect example of the shielding produced by high energy charged particles crossing over the magnetic field lines of the Earth, except of course near the geographic poles. Around the geographic poles of the Earth, the charged particles experience no force, hence deflection, when they move parallel to the Earth's magnetic field lines, since the sine of zero degrees is zero and therefore the magnetic force on them drops to zero. Even at small angles near the poles the force is greatly reduced, so these regions are much more susceptible to the "northern lights" phenomenon. This example is identical in principle to the one we just went through, and is a great illustration of how we are protected by heavy cosmic ray bombardment while forming "belts" of radiation orbiting the Earth. The second example brings us full circle (no pun intended) back to the electron. Since it is negatively charged, we need to flip the rule and use our *left hand rule or LHR for the electron*, but all the conventions are otherwise the same. Thomson and others were able to send cathode rays, in other words a stream of electrons, through a vacuum in a glass tube, and by applying an external magnetic field in such a way that the electrons passed through this field, they would change direction according to the LHR. Ingenious experimental setups, such as Helmholtz Coils, could bend the electrons into a circular path, and using the equivalence of $QvB = mv^2/R$, the charge to mass ration Q/m was then obtained. The speed "v" was ascertained through knowing the voltage used, and the radius R of the path taken was easily measured. Once Millikan did his Oil Drop Experiment and determined the elementary charge on an electron, the only quantity left unknown was the mass of the electron. This is how we finally deduced that its *rest mass,* or mass at speeds less than 10% the speed of light, is 9.11×10^{-31} kilograms. Think about that for a moment: there is no direct way we can measure such a small mass, just as we can never measure directly the mass of the Earth or the Milky Way. And yet with our minds and through experimentation, we know all of these numbers to a remarkable degree of precision. This computer, and all of electronic functioning, rests squarely on our precise knowledge of the mass and charge of an electron.

There is a very clever method to find the speed of these charged particles experimentally by combining an external electric field with the external magnetic field. Going back to our example of the proton moving across the page to the right within a magnetic field that is directed into the page, we discovered that the magnetic force was directed upward in the plane of the page as the proton enters the B-field. What if we used another, downward force to balance this upward magnetic force, causing the net force to be zero and thus sending the charged particle straight through un-deflected? This scheme is dubbed a "velocity selector", because if set up correctly it will do just that: select a particular velocity for the proton to travel straight instead of in a curved path. Gravity is much too weak to use as the downward force on the tiny mass of the proton, so instead we use the electric force. Suppose we put positive charges just above all the X symbols and negative charges just below all the X symbols. This sets up a situation where the electric force on the proton produced by this external field is downward, since like charges repel and opposite charges attract. By fiddling around with some dials, we can adjust the voltage producing this external electric field (E) so that the proton will experience zero net force and pass straight through, and this will only occur at a specific voltage, hence proton speed.

To drill down a bit with this, we need to talk about the EM force on a charged particle that is moving in the presence of both an external electric field and an external magnetic field. This is called the Lorentz (or EM) Force, and the formula looks like the following. Vector quantities will be denoted by using boldface, since the directional nature of all of this is extremely important. Here is the formula: $\mathbf{F}_{EM} = Q\mathbf{E} + Q\mathbf{v} \times \mathbf{B}$. The first term is the electric force on the proton caused by the interaction of its charge with the charges producing the external E field, and the second term you should now recognize is the magnetic force on the proton caused by the magnetic field it produces (since it is moving) interacting with the external B field. This means if we set these two forces equal (and oppositely directed), the total or net force will be zero and the proton will pass through the region in a straight line path. Setting the magnitude of the two forces equal, and not worrying about direction any longer because we made sure the forces oppose each other, we have: $QE = QvB$. Simplifying that we have $v = E/B$. We can measure the E field by adjusting voltage, and we measure the B field in any number of ways experimentally. The result is as simple as it is useful: when the voltage is adjusted just right, the proton will pass through the region in a straight line, and by recording the E and B field magnitudes in that situation, we will know the speed of the proton (which cannot be seen of course and is not otherwise easily ascertained).

You may wonder how we know the proton is going straight, since we cannot see it. There are several ways, but I will list two of them here as examples. We could set up a detector in the way of the straight line path we want the proton to take and keep track of when it starts to register a charged particle hitting it (similar to the Rutherford scattering experimental detection method), or we can run the experiment in a cloud chamber whereby the passing charged particle will leave a visible trail of bubbles. By now you may also have realized that this is how we can use the centripetal force method to establish the mass of any charged particle moving in an external magnetic field, and that is one way we know the mass of a proton!

While writing this piece, it occurred to me that some readers might have done some excellent "connectivity" work, and asked themselves the following question. Why, when we throw a baseball or football horizontally through a gravitational field, does the ball follow a *parabolic* path instead of a circular path? After all, the ball enters the external gravitational field with a velocity perpendicular to the direction of the gravitational field (which is directed downward of course), so why is it that the ball doesn't do what the proton did in the external magnetic field and thus move in a circular path? We might broaden the whole question and ask why the parabola is the preferred path nature chooses over all the possible paths the ball might take. We need to explore this line of reasoning at some depth, because it leads to yet another amazing insight into nature. Once we take in that scenic view, we will close this section by returning to the hydrogen spectrum and unlocking one of the mysteries the Bohr planetary model could not answer.

To begin with our foray into this question of parabolic flight, we need to see that the two situations of the proton and the baseball are not exactly the same. Initially it may seem that way, since the force causing the motion starts out as perpendicular to the object's velocity. But look what happens to the baseball as it moves along in its flight. Its *horizontal* speed and direction are unaffected by gravity, and so both remain constant in the absence of any significant air resistance or wind. The baseball's *vertical* speed, however, is a whole other matter, because now we can see that the force of gravity is going to change (in this case increase) the baseball's *vertical* speed along the flight path, since the force is downward and so is the vertical component of the baseball's velocity. This is not the case with the proton's motion through the external B field. In that case, the force is always perpendicular to the proton's velocity, so its speed cannot change, which results in uniform circular motion.

Why is the path *parabolic* then? I will point you in a fruitful direction only; the rest is up to you and how curious you are about the answer. Look up the Principle of Least Time, or the Principle of Least Action, or Fermat's Principle. Any one of the three should lead you to fertile ground. It is extremely tempting to follow along with you on this sidebar, but I believe it was Feynman who once said (or wrote) that a good teacher knows when to be silent, so I will follow his advice. One final point though, before moving on to the closing section dealing with the hydrogen spectrum. This least time principle is incredibly significant in physics – it shows up in deep and fundamental ways in nearly all branches of the discipline. It really is worthy of your pursuit; I hope you take that path and find out why baseballs move in parabolic flight through a gravitational field.

We cannot leave this section on the EM force without a trip back to the spectrum of hydrogen, and a brief discussion and analysis of the experimental results that were explained in a coarse fashion by the Bohr planetary model of the atom. This in no way is meant to imply that the original work done by Bohr and others, establishing the wavelength of the four visible emission lines (called the Balmer Series if you recall) in the hydrogen spectrum, is somehow not worthy of our admiration. To the contrary, the fact that Bohr worked from some fundamental physics already established by Coulomb, Maxwell, and Newton, and then added to it some important ideas of his own to finally predict where those emission lines came from and what their wavelengths would be, is an accomplishment that rightly earned him the Nobel Prize in physics in 1922. One of those assumptions was certainly an outcome of the quantum world that was just beginning to gain firm footing in physics, namely that only certain energy levels were allowed in the atom, and these energy levels were unique to each element on the periodic table. This is why every element has a unique and identifying spectrum. Therefore, the photons that were emitted when electrons jumped down to a more stable energy level had to have a specific energy that was equal to the difference in those energy levels in the atom, and according to Planck's $E = hf$ assumption in blackbody radiation, that meant the emitted photon had a specific frequency and wavelength that could be predicted. It worked wonderfully, but there were some issues with the model, in that it did not account for several observations *within* those spectral lines. There apparently was more to the story, and I will focus on a few aspects of it, starting at the most logical point available: actual observations and data.

Michelson and Morley, of the "no ether" fame, had developed extremely precise experimental equipment to study the spectrum of hydrogen gas placed under high voltage. The emission spectrum observed looked like a superb match with Bohr's predictions, but upon closer inspection, some spectral lines were actually split into doublets that were extremely close to one another. Instead of the one red line corresponding to the 656 nanometer wavelength photon that was emitted when an electron jumped from the second excited state (n = 3) to the first excited state (n = 2), there were actually *two* red lines closely spaced next to each other. This "fine structure", as it was appropriately called, seemed to imply that some of the energy levels of the electron in the hydrogen atom were actually split into nearly equal multiple levels, thus producing the multiple emission lines that were being observed. What could cause this fine structure within the atom?

The first attempt at an explanation was an excellent one that offered much insight and considerable success. Sommerfeld and others, not long after Bohr did his work, suggested that perhaps the origin of this splitting of spectral lines and energy levels was due to relativistic conditions, namely the increase in electron mass as it approached the speed of light in the atom. Dirac added to this analysis by developing a complete description of the relativistic effects on the energy levels of the hydrogen atom, and came up with an equation which matched the experimental observations exceedingly well. The magnitude of these effects was calculated to be equal to a dimensionless quantity, called the fine structure constant, and the constant that tumbled out of the analysis is as follows:

Alpha = α = fine structure constant = $e^2/(4\pi\varepsilon_o hc/2\pi)$ = .00729... ≈ 1/137 with no units. This is a pure number, and apparently is yet another fundamental constant woven into the fabric of the universe that comes from a combination of knowing the elementary charge on an electron, Coulomb's electrical force constant, Planck's constant, and the speed of light. From quantum electrodynamics (QED) and further progress, we now know this number to a precision that is unmatched just about anywhere else in physics. But look at how remarkable this whole enterprise has become! Our measurement of the fine structure of these spectral lines (hence the atom) starts with Bohr, leads us to relativity, and that is not the end of the story yet. Along the way, appreciate the fact that its value depends on all the work done previously to establish the elementary charge on an electron, the speed of light, the strength of the electric force between charges, and Planck's constant associated with photon energy. The fine structure constant is sometimes referred to as the strength of the interaction between electrons and photons. The atom is clearly an electrical place, and things are whizzing around in there at a high rate of speed apparently. Perhaps you noticed I slipped a little caveat into this explanation a short time ago – the point where I wrote "this is not the end of the story". Read on.

Nowhere have we included the magnetic effects within the atom, and this cannot be completely correct, since we have already established that moving charge produces a magnetic field. We now enter the territory – perhaps we were well within its confines already – where analogies can be wonderfully useful in visualizing processes and yet completely unrelated to the reality of the situation. I will not go into the details extensively here, but we will look at the following points only to emphasize that the atom is actually, more accurately, an *electromagnetic* place. As the electron cloud is established around the nucleus and described by Schrodinger's equation, we know there must be something called orbital angular momentum, and it, too, must be quantized. We will discuss this much further in the last section on the conservation laws, for at that time the concept of angular momentum will be explored in full. For now, we can think of the atom as having orbital angular momentum due to the electron existing in specific orbitals about the nucleus.

In addition, we think of the electron as having a "spin" angular momentum of its own. This is what allows the ground state of an atom, the energy level that is most stable, to have two possible orientations within the same energy level, called "spin up" and "spin down". The Pauli Exclusion Principle thus allows two electron states to occur within the ground state. In one case the spin axis of the electron aligns with the orbital angular momentum axis of the whole atom, and in the other the spin axis is anti-parallel with the atom's angular momentum. If it helps to think of all these things by visualizing an electron as a spinning sphere or top, much like how the Earth rotates on its axis as it also revolves in orbit around the Sun, then by all means do so! But remember, there is no indication that electrons have any real structure at all – they are as close to point particles as it gets. The analogies don't matter if the equations work to predict the spectral lines, and they do!

The next step is the fact that the nucleus has a magnetic spin or "moment" as it is called, something we use very effectively when an MRI is done in an attempt to discover potentially lethal problems in the biology of human beings. Again, much of this seems to be quite abstract, and some of it most definitely is, but the miracle is that the thinking and the math work to explain the workings of nature, and in the case of MRIs it saves lives! The interaction of the electron's magnetic effects with those of the nucleus introduces the next level of splitting, called hyperfine structure. There are also quantum effects with the electron that introduce still more "fine detail" in terms of atomic behavior and spectral analysis. The attempt here is to introduce you to some of the finer points involved with the EM force, and how it affects the atom (and the electrons within the atom). In passing, it is this electron "spin" which we use to identify which elements will be amenable to becoming magnetic, either by putting the sample near a powerful "permanent" magnet, or by running a current through a nearby wire. In magnetic materials such as iron, electron spin lines up with the external magnetic field, creating domains within the sample that give it its magnetic properties.

Nature appears to be like the proverbial ultimate onion: each layer peeled away leads to another deeper, finer layer to be explored and hopefully understood. Along those lines, I agree with the notion, expressed by several physicists and others, that understanding what makes a rainbow work the way it does has no effect on the ability of a person to appreciate the beauty of that rainbow. In fact, I would argue that knowing the intricacies of how that rainbow is formed only adds to the mystery of the workings of nature and nurtures a deep appreciation for the complexity and splendid colors in its display. I believe an abiding respect, indeed reverence, for our Earth and our universe, is more likely when one begins to establish a deep connection with the patterns we are all free to observe and wonder about. When I come across the number 137, or the speed of light, or pi, or any such constant of deep significance, the sense of excitement, passion, and power is not unlike hearing a piece of music that moves the spirit in a manner words cannot. The same truth holds when I am *seeing*, not just looking, at a Van Gogh masterpiece. I think all such experiences are important for making the most of our short existence between two eternities. The last section of this book will delve into the four fundamental conservation laws that serve as the pillars, the cornerstones of this edifice we call physics. Each one rests on deep symmetries in nature that we still do not fully comprehend.

Four Conservation Laws: Linear Momentum, Angular Momentum, Charge, Total Energy

The importance of conservation laws has gradually gained steam over the past century or so, and at the present moment they occupy center stage in the physics world. The reason for this is straightforward: *every single piece of experimental evidence verifies and confirms the validity of these laws, without exception, over every field of physics and across all the sciences.* What does it mean when it is said or written that something measurable in physics is conserved? Quantitatively, a conservation law is telling us that in a closed system with all factors taken into account, the total amount of the item being conserved remains constant throughout the process. We will see the quantum world allows for a tiny bit of leeway here during certain processes, but in the end the conserved item under study remains unchanged. From a conceptual and qualitative viewpoint, the deeper meaning laying at the heart of all conservation laws is that these are absolutely dependable rules of nature because they rest on beautiful symmetries inherent in our universe. With regard to problem solving, the conservation laws often unlock puzzles in situations where all other methods either fail or incur complexities that needlessly block solutions while also consuming large amounts of time and energy in the process. This metaphorical climb is our last pitch to reach the summit of the book. Again, I hope the reader, (and the author) will keep the focus on each step of the journey rather than the end result.

CONSERVATION OF LINEAR MOMENTUM

As we encountered many pages ago, Newton developed a deep insight into the natural laws that govern all processes in the universe when he proposed that for every action force, there is an equal and opposite reaction force. This suggests that if two people standing on frictionless ice push off one another, each will gain velocity but in opposite directions. Which person will gain more velocity? We could use another scenario in which two astronauts are stationary (relative to each other) in outer space, and one of them decides to toss a toolbox to the other astronaut. Assuming that they are risk takers and neither of them is tethered in any way, what would happen in such a scenario?

Initial analysis regarding the situation on frictionless ice would indicate that the person with the lesser mass must gain more velocity than the person with greater mass, since the force between them is one (and the same) interaction. Exerting this mutual push through a time interval will result in a greater change in velocity for the lesser mass. For the astronauts, things are a bit more interesting. Again, we assume no ropes or strings are attached and there are no external forces unaccounted for in this "closed system". When the first astronaut throws the toolbox toward the other astronaut, he must recoil backward according to the third law of action-reaction. When the other astronaut catches the toolbox, she (and the toolbox) must move away from the receding male astronaut. Now if she were to throw the toolbox back to him at this point, we can see there is a dire situation, because the reaction will cause her to move even faster away from him, and when he catches the toolbox this simply adds to his motion away from her. Before long, they will both be lost in space! How can we quantify these things, so that we can predict what the velocities will be in such situations?

We start with Newton's Second Law, and I will use boldface again at this point to denote vector quantities, because the topic under investigation relies heavily on direction. Recall that this law states the net force (sum of all forces) acting on the object must equal its mass times acceleration, or in terms of a formula: $\mathbf{F}_{net} = \mathbf{ma}$. Suppose we multiply both sides of this equation by the time of interaction between any two objects, such as described in either of the scenarios, and recognizing that the force applied must be the same for both by Newton's Third Law of action-reaction. This gives us: $\mathbf{F}_{net} \cdot t = \mathbf{ma} \cdot t$. Now we use the definition of acceleration, which is a change in velocity divided by time: $\mathbf{a} = (\mathbf{v}_f - \mathbf{v}_0)/t$. We substitute this expression for the acceleration back into the first step, and we are left with: $\mathbf{F}_{net} \cdot t = m(\mathbf{v}_f - \mathbf{v}_0)$. This must mean that since the force and the time have to be the same for each object, as this is the same interaction they each experience, the quantity on the right side of the equation must also be equal (and opposite) for each object. Whatever we call this quantity, it then becomes clear that whatever one object gains of it in one direction, the other object must gain the same amount in the opposite direction. *Therefore, the total amount of change in this quantity must always be zero in a closed system!* The coda to this story is simple: we give this quantity of mass times velocity the name "linear momentum", and represent it with the letter **p** (since m is already taken as the symbol for mass). Hence, in general, **p** = m**v**. We then conclude that in any closed system, the total amount of momentum, *in each direction independently since it is a vector*, must be conserved – in other words, it must remain constant.

At first glance we can see that this is an alternative way to look at the concept of force: we could have defined force as the rate of change of momentum, and this is what Newton and many others did. The units for momentum are kilograms x meters/second, and there is no name or label yet (at this writing) for momentum units, such as there are for force, energy, power, and so many other physics concepts. Perhaps this is a good thing, because it forces us to see that momentum is a unique combination of mass, space, and time. The fact that this combination is conserved in a closed system has immense consequences in our universe, and provides an incredibly powerful tool in problem solving. Illustrations are worth far more than abstract notions at times, so let's look at a few examples to see how this conservation principle can unlock numerous puzzles in nature.

Suppose the engine of a train is moving due East on a set of railroad tracks at a speed of 10 mph when it collides with a stationary railroad car resting on the same track. We don't know the masses of the two objects directly, but we do know that the engine has four times the mass of the railroad car. When they collide, what will the final result be? Using the conservation of linear momentum, we can set this up as follows. Take the mass of the railroad car as "m"; this means the mass of the engine must be "4m". The initial momentum of the closed system is then given by adding the m**v** contribution of each mass: (4m • 10) + 0. The first term is the engine moving at 10 mph and the second term is the zero momentum of the railroad car since it is not moving initially. The final momentum of the system is the combined mass (since we assume they stick together upon collision) times the unknown velocity we are trying to predict. In equation form, the conservation law would look like this: $4m \cdot 10 + 0 = (4m + m)v$. Simplifying, $40m = 5m(v)$, which we solve immediately by canceling the "m" and getting $v = 8$ mph! It makes sense that the final velocity would be less than the original 10 mph, because we have added mass to the system during the collision. Also, since all the motion is in one direction, it makes sense the final velocity must be East also.

Here is another example, and we will use the astronauts for this one, a man and a woman. Suppose he has a mass of 100 kilograms, and he tosses the toolbox of mass 50 kilograms to the right at a velocity of 4 meters/ second toward the woman. For the first step, the initial total momentum is zero since both the astronaut and the toolbox are at rest. This means the final total momentum must be zero as well! In equation form (we will take "right" to be in the positive direction here):

$0 = 50(4) + 100v$ ... solving for the velocity of the astronaut we get -2 meters/second. The astronaut has gained -200 units of momentum toward the *left* (hence the minus sign), and the toolbox gained 200 units to the right, so the total change in momentum of the closed system is zero, as it must always be since momentum is conserved. Notice we are looking only at systems which move along straight lines here – hence *linear* momenta are involved. We will deal with angular momentum in the next part of the journey.

Whenever a process involves collisions or explosions, the conservation of linear momentum shows its remarkable problem solving power. Actually, one could think of an explosion as just a collision in reverse time, so perhaps it is not surprising that the same principle applies equally to each situation. What if the two objects are moving straight but not along the same line of action? Total momentum is still conserved, and in this situation we make sure to conserve the momentum in each direction independently. Here is one example, chosen in such a way as to keep the math to a minimum and also to appeal to your innate sense of symmetry (something we will come back to shortly). Suppose a truck is moving due South along a highway toward an intersection, traveling at a speed of 30 mph. At the same time, a car is moving due East along another highway toward the same intersection at some unknown speed. Unhappily, they collide in the intersection, and their fenders lock so that the total mass of both ends up directly Southeast (45 degrees from both the East and South axis). The only thing we know for certain is that the truck has three times the mass of the car. Which driver is far more likely to be at fault? Yes, physics is often employed in crashes (and explosions) of all kinds in order to determine what events happened prior to the accident. This is one powerful tool we can often use to reconstruct tragic events such as plane crashes with the hope of preventing the same thing from happening in the future.

I have intentionally given you some time here to ponder the answer to our collision and figure out who is at fault. Momentum must be conserved in each direction independently, so that means the initial South momentum is only the truck (since the car was moving East), and that is equal to 3m • 30 = 90m, where "m" is the mass of the car and "3m" is the mass of the truck. The initial momentum in the East direction must therefore be m • v, the mass of the car times its (unknown) velocity. Now for the key fact: the wreckage was found to be at a 45 degree angle from the point of collision, so that must mean each vehicle had the same *magnitude* of *momentum* in each direction *before* the collision, otherwise the final wreckage would have favored one direction over the other! Therefore, since the original momentum was 90m for the truck, the car must have had the same *magnitude* of momentum, and hence the car must have been traveling at 90 mph! I am guessing many of you came to this conclusion already, perhaps without using any equations at all. That is because we might have had experiences, such as by playing football or in some other contact sport, that have "taught" us what happens. On the other hand, it might be we have an innate sense of symmetry. Whatever the case, the conservation of linear momentum has unlocked this particular puzzle.

Returning to the original formulation where we had $\mathbf{F}_{net} \cdot t = m(\mathbf{v}_f - \mathbf{v}_0)$, we call the left side of this equation "impulse" and the right side of the equation "change in momentum". There are literally millions of practical applications that flow from this relationship. Why do we follow-through in baseball, golf, tennis, and so many other sports? Because it increases the time of contact, which increases the impulse, which gives the ball a greater change in momentum to go either farther or faster or both. Recall the karate expert who is able to split a concrete block: by keeping the time of contact very small, the force exerted must be quite large. Also, by reducing the contact area of the hand on the block, the pressure (force/area) exerted on the block can be enormous. Another example is rocketry, in which high velocity fuel ejected out the back of the rocket must propel it forward. One hopefully begins to see how the action-reaction law is intimately tied to momentum conservation.

Total linear momentum conservation has proven to be especially powerful when we analyze atomic and nuclear processes. I will present three examples here, but there are many millions to be sure. First, when Chadwick was doing his experimental work in the 1930s, it had been standard practice for years, ever since the Rutherford (Geiger and Marsden) gold foil scattering experiment, to bombard nuclei with alpha particles (helium nuclei remember) and see what happens after the "collisions" with the positive nuclei. It was noted by several physicists, *based on the conservation laws of linear momentum and total energy*, that the product of some of these collisions appeared to be highly energetic neutral particles. Some made the conjecture that perhaps this penetrating radiation coming out of the collisions was gamma radiation, but further analysis showed that this could not be the case. Chadwick was well aware of the ongoing work, and decided to bombard the element beryllium with alpha particles (what else?!) produced from radium samples. By analyzing these results using momentum and energy conservation, he was able to prove that the highly penetrating, neutral particles coming out of the collisions were, in fact, neutrons with about the same mass as the proton. This solidified the conception of the nucleus being composed of protons and neutrons, and we have already seen the many places that led us to in the ensuing years. Chadwick won the Nobel Prize for this work in 1935.

The second example is a wonderful illustration of how confusing the world of quantum physics was (and still is, in the minds of many!) during the first half of the 20th century. Arthur Compton got the idea of scattering photons off electrons, and predicting what might happen as a result. Again, this is a collision on the subatomic scale, and therefore momentum conservation becomes a very useful tool. Compton treated the collision much like that of a billiard ball colliding with another, initially stationary, billiard ball. Therefore, you might begin to comprehend the confusion, since much of the work prior to this had shown that electrons had wave characteristics (Davisson-Germer experiment is one example of this), and here was Compton treating electrons like particles akin to billiard balls. Of course, the wave-particle duality of light (photons) was already hundreds of years in the making, so there emerged the tongue-in-cheek witticisms, such as "wavicles" to express *this* apparent contradiction. The bottom line to all of this is that Compton's treatment, based on the conservation laws of energy and linear momentum, predicted the experimental results almost perfectly. The scattered photon gave some of its energy up to the electron upon collision, and moved off at various angles with a

"Compton-shifted" longer wavelength (less frequency, less energy by Planck's $E = hf$), and the electrons scattered off in various ways from the photons. Again, every one of these scattering results confirmed the conservation laws!

The third example is a personal favorite, and it brings us back to the decay of a free neutron that we explored many pages ago, vis a vis the weak interaction. Recall that in beta minus decay, the neutron decays into a proton, a high speed electron (the beta particle), and the anti-neutrino. The problem was, nobody could detect this ghostly anti-neutrino. In the parlance of experimental physics, it just wasn't there, at least in any way that was known or could be observed. Yet what the results did show, and very clearly, was that it appeared both energy and linear momentum were not conserved in this vital nuclear process. Physicists were faced with a choice: either accept that there are exceptions to the conservation laws, or hypothesize the existence of a particle that would account for the missing energy and momentum apparent in beta minus decay. We chose the latter option, essentially unwilling to consider the prospect of such fundamental laws having significant "exceptions to the rule". Alas, in what one might construe as a major tribute to our belief that the conservation laws are inviolable, experimental evidence for the neutrino and its anti-particle mate was finally uncovered two decades later, and they have been a subject of intense study and research ever since.

The question arises as to what happens if more than two objects are interacting in a system, as is so often the case. In just one mole of a gas, for example, there are over 10^{23} atoms or molecules. How will we follow each one of these particles and trace the collisions they make with each other over time? There is no practical way, at least at this writing, to ascertain the position and momentum of every particle in a gas at a given moment in time, and from there predict all future outcomes. At the macroscopic level, we can make assumptions, several of which are based on Newtonian Mechanics, and from those assumptions and the conservation of linear momentum, we can deduce the Ideal Gas Law: PV = nRT, where P = pressure, V = volume, n = number of moles, R = the gas constant (related to Boltzmann's constant and Avogadro's number), and T = temperature in Kelvin. Moreover, if we look at systems with huge numbers of particles or objects, we can use the concept of center-of-mass to follow the large scale behavior of the system. For example, rather than try to plot the paths of fifty planes or birds flying in some sort of formation, we can find the system's center-of-mass and follow that point, knowing that where the center-of-mass goes, so goes the "flock". The center-of-mass of a system is usually fairly easy to find, as it is a type of weighted average that locates the point mathematically. The salient feature here is the following: *the velocity of the center-of-mass point does not change throughout any collision or explosion*! This method can be employed to describe the motion of any system of particles, large or small, but it is particularly useful when trying to analyze many-particle systems.

The fundamental question remains. *Why* is linear momentum conserved? Is there some type of underlying principle at work here, causing this mass times velocity product to be so deeply embedded into the laws of nature? Physics is often better at answering the "how" rather than the "why", but in this case there is a symmetry which leads to the conservation of momentum. The symmetry is woven into the fabric of space itself, and it shows that if we take our experiment, whether it be beta decay or two colliding railroad cars, and we translate the whole setup through space to another location, the results of the experiment will be identical. In brief, we are saying that the conservation of linear momentum cannot depend on where an experiment is conducted to establish its validity. Fundamental laws of physics should not vary by location. Let's use an analogy, and relate this back to that giant chess game nature plays with us, as Feynman envisioned. At my table, bishops can only move diagonally. This rule should apply exactly the same way if we move the chess game to your house. All experimental evidence to date indicates that this, in fact, is how nature operates. We should thank her profusely for this, because otherwise it would be an extremely chaotic world indeed. In all probability, if such were not the case, the universe as we know it could not exist at all. A brilliant mathematician, someone I doubt many people have heard about, developed the mathematics behind these symmetry based laws, and her name is Emmy Noether. She was able to show that every conservation law has its roots in a symmetry that is pervasive throughout the universe.

One final item before moving on to the next conservation law. One of the major underpinnings of quantum physics is the Heisenberg Uncertainty Principle (HUP), a concept we have discussed briefly a while ago. This principle states that we cannot know *both* the linear momentum and the position of any particle *simultaneously* to an unlimited degree of accuracy. As a formula, the HUP looks like this: $\Delta \mathbf{p} \cdot \Delta \mathbf{x} \geq h/4\pi$. Experimentally, this means the more precisely we can nail down the momentum of a particle, the less precise our measurement of its position becomes. Again, this is not due to poor experimental work, it is a limiting factor built into nature herself. In the macroscopic world of trees and human beings, this principle is practically impossible to observe, since if we bounce photons off a large object, we can locate its position without changing its momentum by an appreciable amount: Planck's constant "h" is an extremely small number! Yet if we want to locate an electron, think about how that has to be done. When a photon smashes into it so we can say where it is, we invariably cause its momentum to change markedly, by an amount at least as much as $h/4\pi$. The HUP in turn implies that if we cannot know the momentum of a particle exactly, then this conservation law can be "negotiated" over a small extent of space if that space itself is not invariant to the laws of physics. As we have no evidence of any such invariance that I am aware of, I think it is safe to conclude that the conservation of linear momentum is valid at all levels of experimentation, including the quantum world. We move now from linear to rotating systems, and knowing the predictable analogs between them, we would expect that angular momentum is conserved also.

CONSERVATION OF ANGULAR MOMENTUM

Have you ever wondered why helicopters have two propellers, usually oriented perpendicular to each other? How do figure skaters spin so rapidly, and why does this happen when their arms are drawn inward toward their bodies? From our analogies drawn between linear and angular motion, we discovered that the physics equations are exactly the same, but the symbols used are different. For linear dynamics, Newton gave us **F_{net}** = m**a**. The rotational analog to that law was easily translated. Instead of net force causing linear motion through space with changing linear velocity, we have net torque, causing a rotational motion about an axis in space with changing angular velocity. Instead of mass, we have the distribution of mass in relation to the axis of rotation, called the "moment of inertia" (I). And of course instead of linear acceleration we have angular acceleration, with the final result being τ_{net} = Iα. The same analogy should hold for momentum: linear momentum = **p** = m**v**, so angular momentum must be equal to the moment of inertia times angular velocity. In equation form, this looks like the following: **L** = Iω, where "L" is the angular momentum and the other two symbols we already know. This result can be obtained in exactly the same way that we derived the linear momentum formula: multiply both sides of the net torque equation by time and simplify!

Notice that angular momentum is a vector: it has direction because the angular velocity is either clockwise or counterclockwise, about the axis of rotation, so we say that the object is either "spin up" or "spin down", just like we did for the electrons. There is an alternative expression for angular momentum that applies nicely to an object of mass "m" traveling around in an orbit of radius "R" with a velocity "v": **L = mv x R**. When two vectors are multiplied together, we call that a "vector cross product", and there are special mathematical techniques used to evaluate the product. If the velocity and the radius vectors are always perpendicular to each other, such as with circular motion, then the magnitude of the angular momentum is simply L = mvR. The applications here are quite obvious: electrons orbiting the nucleus, planets orbiting the Sun, and many more. We will return to some of those in this section.

There is something that I am sure you know but I am less certain you may have thought about. Why is a bicycle so much easier to balance when it is moving, versus when it is stationary? Take a bicycle wheel off the frame of a bike, or any object that you can spin about an axis, and set it into rapid rotation using an external torque (a few tangential tugs on the rim will do the trick). Now carefully place one end of the axle on a fingertip, with the axle running parallel to the ground (orient it horizontally). Now let the wheel go. It doesn't fall like it would if it was not spinning! Instead, it starts to "precess" around your body, so you have to slowly spin in a circle to keep up with the whole wheel. Now hold the spinning wheel with one hand gripping the end of the axle. Move the axle back and forth, parallel to the ground or floor. Nothing much happens. Now try to point the axle vertically upward by swiftly rotating the spinning wheel so that the axle points to the sky. Suddenly your arm (and the wheel) gets yanked to the right or left by some mysterious torque. Point the axle down quickly and your arm gets yanked in the opposite direction. When you try to move the wheel's axle to the right of left, the torque produced yanks your arm either up or down!

The physics principles behind these phenomena can become quite complex, but the conceptual framework is straightforward. Once we establish the angular momentum of a spinning object, the law of angular momentum conservation requires that the *direction* of this momentum remain the same. This is the vector nature of momentum, and it is the reason spinning bicycle wheels make it easier to balance: nature fights any change in the axis orientation, so the bike whizzes along. This is also why we spiral a football: it is much more stable in flight because the axis of rotation does not want to change its orientation. The same applies to spinning a satellite, or a bullet from a rifle. When we *do* forcibly change the axis direction by moving the axle up or down, left or right, an external torque is caused because we are changing the angular momentum's direction, and this torque is always at right angles to the original motion being executed. Moving the axle in and out, parallel to the floor, causes no change in the angular momentum direction, hence no external torque ensues.

Why does the bicycle wheel precess when it is balanced on a fingertip while spinning? The answer is subtle, but only because *we* are not doing anything to the wheel. Instead, gravity is pulling the wheel down, creating a torque about the pivot point of your fingertip, and that causes the precession (right or left depending on the spin direction of the wheel)! It's the same effect as the person physically trying to point the axle downward while the wheel is spinning. There are several marvelous examples and applications that flow from the directional aspect of angular momentum conservation. Let's look at these, starting with helicopter flight dynamics.

All conservation laws imply that whatever is being conserved cannot be created from nothing in a closed system. As one concrete example, sit on a stool that is free to rotate, and without touching anything external (so as to impose the closed system requirement), try to spin the stool continuously by twisting your body. You will note that your upper body spins one way, while the lower spins the other way: you can see this when a shortstop makes a great backhand stop and leaps in the air to throw out the runner. No matter what you do, assuming the floor is level and you use nothing external, that stool will not spin continuously. Angular momentum cannot be created from nothing! Now we turn our attention to the helicopter. The main rotor spins in one direction, thus establishing a great deal of angular momentum when up to speed. As soon as the helicopter lifts off the ground, the Earth no longer can "absorb" this spinning, and the helicopter itself starts to spin in the other direction to conserve angular momentum. Therefore, a rear "stabilizer" propeller is mounted perpendicular to the main rotor, and acts to fight this tendency of the helicopter to rotate opposite the spin of the main rotor. This is why any loss of the rear propeller during flight is catastrophic, as the helicopter will then spin out of control. Very large helicopters typically have two main rotors, and of course they are made to spin in opposite directions, synchronized in such a way as to produce stable flight.

Another great example, and application of these principles, occurs in racing cars held on oval tracks. Most cars are engineered so that the crankshaft, a spinning metal pipe running the length of the vehicle, transfers the energy from the engine to the flywheel and thus to the wheels, and when viewed from behind the car, the flywheel spins counterclockwise. This means whenever the car turns a left-hand corner, the effect is to produce an external torque that draws the car down, into the road, creating better stability through traction and a lower center of gravity. Though there are many reasons for races of all kinds to be run counterclockwise, and car races can also run clockwise, there is no question that these gyroscopic effects of high speed vehicles are critical factors in developing stability in the engineering process. High speed, precision gyros are used throughout all facets of flight dynamics, from directional gyros used in automatic pilot systems to tremendously sophisticated inertial navigation systems.

The above discussion is a brief foray into the directional nature of angular momentum, and how this relates to its conservation principle. We now investigate the other aspect of the angular momentum vector: its magnitude. Returning to the figure skater, when she draws her arms inward from a slow spin rate, she is reducing her moment of inertia, because now her mass distribution is closer to the axis of rotation, making it easier to spin. Since the magnitude of angular momentum equals the product of the moment of inertia times angular speed, and angular momentum must always be conserved in an isolated or closed system, if the moment of inertia decreases, the angular speed must increase. This is what causes her to spin much faster! The "governor" in a small engine, such as might be used in a golf cart, works in a similar fashion. If the vehicle starts to move too fast, the "governor" is a device that spreads mechanical arms outward (or uses some other engineered system) at higher speeds, thus increasing its moment of inertia. Since angular momentum must be conserved, this ends up decreasing the angular speed of the device, which can be linked in some fashion to controlling the speed of the cart.

Now we are in a better position to move from the largest masses we know in the universe, down to the smallest, and come at both from the perspective of angular momentum conservation. Two black holes orbiting each other will start to spin around faster and faster as they approach one another, due to less system moment of inertia, hence greater angular speed. It always gets back to $\mathbf{L} = \mathbf{I}\boldsymbol{\omega}$: since $\mathbf{L}$ must remain constant, if I goes up then $\boldsymbol{\omega}$ must go down and vice versa. Therefore, when the two black holes do their primordial dance and crash into one another, the gravitational ripples in spacetime produced will have that increasing frequency as measured by the LIGO systems we have explored. Likewise, neutron stars are formed by massive collapsed stars, and since the star was spinning initially, just as our Sun does, the collapsed mass will have lesser moment of inertia and greater angular speed. Some of these neutron stars are spinning at extremely rapid rates (for a star), down to milliseconds per rotation! In doing so, they send out exceedingly precise "pulses" of EM radiation as seen by observers on Earth, hence the term "pulsars". Moving down in mass to planets, the Earth precesses as it spins on its axis, or "wobbles" like a spinning top, because it is not a perfectly uniform, solid sphere. Due to its spin, the Earth bulges at the equator and flattens somewhat at the poles, creating an object called an oblate spheroid. Therefore, our planet's rotational axis traces out a circle, completing one precession every 26,000 years. Yes, that does mean every 13,000 years in the cycle, seasons flip in the two hemispheres, since the Earth is tilted on its axis relative to the Sun. Also, that implies that the length of the solar cycle is increasing ever so slightly, since more mass away from the axis of rotation means a larger moment of inertia, hence slower angular speed. This was one of the compelling reasons for moving to much more precise cesium clocks – the solar cycle is not constant (enough) for the precision we now require.

We have seen how this principle of angular momentum conservation is everywhere around us in "ordinary" life, whether it be in cars, bicycles, helicopters, figure skaters, sports, or in the navigational systems of satellites and rockets. Naturally, we end this branch of the journey by looking at the smallest of items: the atom. We are going to approach this by combining a whole slew of concepts, and in the process use some heuristic arguments to come up with a few startling results. The chain of logic has a starting point at the wave-particle duality of *all* things as postulated by Prince Louis de Broglie. His contribution was that the wavelength of all matter is equal to Planck's constant divided by the momentum of the object in question. As a formula, this looks like the following: $\lambda = h/mv$, where λ is the wavelength and the remainder of the symbols are now somewhat familiar to the reader, hopefully. This is a fascinating idea, so I will first apply it to the macroscopic world of human beings, cars, and so forth.

A quick calculation will show that the wavelength of a human being is incredibly small, since Planck's constant "h" is of the order 10^{-34}. In fact, the wavelength is so small that we cannot diffract or interfere through any set of slits, such as light waves do in Young's

Experiment; instead, we just pass through intact as particles do. In general, the smaller the wavelength, the more the object behaves like a particle. An electron, on the other hand, has a rest mass on the order of 10^{-31} kilograms, so already we can see the de Broglie wavelength will be much bigger than for humans. Even at high speeds, the wavelength of an electron is around the size of an atom, in the vicinity of 10^{-10} meters. This means if we could send electrons through a slit of that magnitude, we should be able to see a diffraction pattern, as we do for any wave. And the miracle is, we *can* do this, and we have, many times. By sending electrons through the lattice structure of some materials, we observe the diffraction pattern produced by the electrons. The layers of the lattice structure are separated by very small distances close to the size of an atom, and those spacings provide the slit widths necessary for the diffraction to be observed. Davisson and Germer were the first to do this, using nickel crystals in the 1920s.

The second piece of the puzzle rests on some of the assumptions Bohr made when predicting the spectrum of hydrogen using quantized energy levels. Again, these assumptions, or postulates as they were called, do not take into account relativity theory, and are proposed ad hoc. If electrons have wavelengths, then to be stable in an orbit, it is expected there will be resonances where the wavelengths constructively interfere with themselves. This would occur if an integral number of wavelengths of the electron fits perfectly into the circumference of the circle that it travels in, since in doing so each orbit will be in phase with the next one. In mathematical terms, we are suggesting that "n" wavelengths must be equal to $2\pi R$, where n is any integer 1, 2, 3, and so on. The formula looks like this then: $n\lambda = (2\pi R)$.

Now for the third piece of the puzzle. We equate the two expressions for the wavelength of the electron: $h/mv = (2\pi R)/n$. Rearranging terms a bit so we don't have to look at fractions, we cross-multiply and get: nh = 2πmvR. Perhaps something is popping out at you at this point, because the "mvR" piece is looking awfully familiar. It is the angular momentum of the electron! Now we can bring the 2π to the other side of the equation, and we have nh/2π = L. This means Bohr was saying, even though he did not come at it this way, since de Broglie did his work *after* Bohr (and in fact used some of this work to develop his wavelength), that *the angular momentum of an electron in the hydrogen atom is quantized*, along with the quantized energy levels of the electron in the atom. Once again, the quantum world appears to operate in discrete "chunks", and is not continuous like so much of the macroscopic world appears to be.

I cannot close this section without exploring the de Broglie hypothesis a bit further, and demonstrate how it can lead to some very famous relationships, albeit along erroneous paths fraught with poor assumptions. It is of lasting interest to me that even though Bohr made many such assumptions based on shaky and sometimes incorrect reasoning, the end result was a prediction of spectral lines that was a phenomenal success at the time, despite the shortcomings we have discussed (fine structure, varying intensities, and so forth). First, if we rearrange de Broglie's matter wave hypothesis by multiplying both sides by the momentum, we have: mv x λ = h. This looks suspiciously like the Heisenberg Uncertainty Principle! If we take the uncertainty in the electron's position as the wavelength, which seems reasonable since the electron is somewhere or everywhere along that wavelength, then we have something that resembles the expression $\Delta \mathbf{p} \cdot \Delta \mathbf{x} \geq h/4\pi$. Once again, this approach is emphatically *not* rigorous, but is intended to illustrate how these ideas might connect conceptually.

The second formula which pops out of these ideas is one you may recognize. We have the energy of a photon from Planck's relation: E = hf = hc/λ (since the speed of a wave is its frequency times its wavelength, or $c = f\lambda$). Next, we take de Broglie's wavelength and recall it is equal to h/mv. Substitute that expression back into the energy formula and we have: $E = hc/(h/mv)$. We simplify that and obtain $E = mcv$ ... and if we take the speed v to be the speed of light c, we arrive at $E = mc^2$. Please make sure to note all the false assumptions that go into this sort of "derivation". We bounced back and forth between an electron that has a nonzero rest mass, and then used a photon which has zero rest mass. That is just the start of some very shaky assumptions along the path to the formula. The actual derivation of $E = mc^2$ takes a much different road in every sense: historically, mathematically, and in terms of physics principles. The notion I leave you with is how even shaky and wrong assumptions can sometimes lead to a result that is interesting, and in the final analysis, revolutionary. I end with the symmetry in the universe that this conservation principle rests on, and it is the fact that the results of any experiment we can do will yield the same results, regardless of how we rotate the setup through space. We call this property of space "isotropic". We move on now to the third leg of the conservation laws.

CONSERVATION OF ELECTRIC CHARGE

This conservation law is simultaneously quite simple, in a physical sense, while also being complex, in that it is difficult to prove mathematically from the standpoint of symmetry. Total charge conservation means that in any physical process within a closed system, the total amount of electric charge must remain constant. Since charge is a scalar quantity, this implies that all we need to do is to count up the charge at the start of the process and set that number equal to the total amount of charge at the end of the process. Like all conservation laws, it is applicable to *every* process in nature, so it has tremendous predictive power. As one example, we can predict that an electron can never decay into a photon, because that would violate charge conservation: a negative one elementary charge cannot end up as a zero charge photon. In the beta minus decay we studied, the neutron turns into a proton, electron (called a beta particle here), and an antineutrino. We start with zero net charge, and therefore we must end with zero net charge, which we do because the proton has a charge of +1 elementary charge, the electron -1 elementary charge, and the antineutrino has zero charge. Yet another example of charge conservation at the subatomic level is electron-positron pair annihilation. When these two particles meet, one positive (positron) and one negative (electron), the result is two gamma-ray photons with zero charge. This process is also a striking confirmation of the conservation principles related to linear momentum and total energy. The conservation of electric charge principle guides physicists extremely well in determining what particle processes are forbidden and thus never occur, and when certain processes do occur, it is found the electric charge is conserved, without exception.

This conservation law also implies that the charge on a proton must be *exactly* equal to the charge on an electron, since if it was not, it would involve creating or destroying electric charge from nothing. Put another way, if they were not exactly equal, then atoms would not be neutral, they would be constantly charged one way or the other. This is not the case, so we conclude that the magnitude of the charge on a proton exactly matches that of the electron. It is a type of indirect proof, where we assume the opposite premise is true, and then proceed to demonstrate that such a premise leads to an absurd or inconsistent conclusion, hence the premise must be false.

It also appears from this analysis that the total amount of charge in the entire universe is constant and has been since the beginning of time. When we rub a rubber rod with wool, we are not creating charge, we are just transferring electrons from the wool to the rod, leaving the wool positively charged for a while, and the rod negatively charged by exactly the same amount that left the wool, assuming no leakage of charge occurs either way, since we invoke the closed system. Even if some charge leaks away into the surroundings, as it invariably does in the experimental process, we simply make our closed system bigger, capturing the leaked charge along with it so that the total charge is completely accounted for. It doesn't take long before we realize if we extrapolate this process, eventually the closed system encompasses the whole cosmos.

Aside from giving us potent predictive power in particle physics, the conservation of charge is also quite handy when it comes to solving electric circuits. We have seen that Ohm's Law works in many situations that involve current and voltage in a circuit, but there are many important exceptions to the law which limit its effectiveness. The conservation of charge has no such limitations, and *always* works, no matter the process. This is where Kirchhoff's Rules enter the picture, as posed by Gustav Kirchhoff in the middle of the nineteenth century. Since electric current is defined as charge divided by time, and charge is conserved, he suggested that *the sum of all the currents entering a junction in a circuit must equal the sum of all the currents leaving that junction. This is called his point rule, or junction rule.* It is tantamount to saying if 30 gallons of water enter a "T" joint in a plumbing section where there is an intersection, and 18 gallons go one way, then the other 12 gallons must go the other way. The total must be 30 gallons, assuming no leakage. This point rule applies to all circuits, without exception, because it is based on a conservation law that always works. When applied to a complex circuit, it becomes indispensable in figuring out the currents in each branch, which in turn gives us a means to solve many other aspects of the circuit.

At the start of this short climb through the conservation of charge principle, I wrote about the symmetry at the root of this particular law, and how it finds its origins in some fairly complex mathematics. The method used to show how charge must be conserved is dubbed "gauge invariance", and it brings us back to Emmy Noether's amazing work in this field, and something called "Noether's Theorem". As that discussion is too involved for this book, I leave it to the reader to pursue. For now, you may have noticed that Kirchhoff coined more than one rule, and that brings us to the fourth and final conservation law we will study, as well as the closing section of the book. While there are more than four conservation laws, we end with what is arguably the most important principle in all of physics, and perhaps all of science: the conservation of total energy.

CONSERVATION OF TOTAL ENERGY

It is fitting that we should close this book with a thorough exploration of a concept extremely hard to define, and yet remarkably simple to employ when unraveling nature's mysteries and patterns: total energy conservation. By now the reader has become acutely aware that words in common usage throughout the world often have precise, defined meaning in the field of physics. Examples include such terms as momentum, force, and power. When the word "energy" is used in physics, the meaning is nailed down completely using the precise language of mathematics, which in turn allows us to form the most important cornerstone in the foundation of physics. You may have come across definitions of energy that read something like "energy is the ability to do work". While this is not wrong, it is significantly limited in its application to a wide array of observations and experimental results. After 40 years working in this field, I have come to the conclusion that there is no perfect definition of energy. The corollary to this might be that I am not sure it matters what combination of words we use to define it, as long as we agree how to *measure* it.

Along the same line of reasoning, I have used the words "law" and "principle" somewhat interchangeably throughout the book. The conservation law of linear momentum, for example, is sometimes referred to as the momentum conservation principle. If you are hearing a tiny sound right about now, so am I: it's the sound of splitting hairs. I will let others exert large amounts of *energy* delineating the difference between the two terms; my goal is for you to *understand* the concept of conservation and, equally as important, how it is *used*! It reminds me of a Zen master who once wrote that when one stops the wind in order to define it, at that moment it ceases to be wind. If that resembles another koan too convoluted to understand, I can assure you that the total energy conservation law is nothing of the kind. We use it all the time, we predict millions of things from it, and it has never let us down. The following demonstration is proof that I have complete confidence in this law, since if it is found to be violated at any time, physics class is immediately canceled while I figure out what happened and thus win the Nobel Prize.

This is one of my favorite demonstrations, used across the globe in the physics world and probably elsewhere, and it involves a simple pendulum with a large mass attached at the end of a wire, with the wire suspended from a ceiling. The larger the mass at the end of the wire, or pendulum "bob" as it is called, the better the demonstration. I have used a 16-pound bowling ball with a hook embedded inside, attached to a steel wire and tied off on an I-beam running across a lecture hall, and at other times just an ordinary mass hanging off a string. Taking the bowling ball, I walk away from its equilibrium position until the wire makes at least a 30 degree angle to the vertical. Drawing the ball up to my chin, I let it go. The massive bowling ball swings downward in its arc, gathering considerable speed, then starts to slow down and finally stops for an instant as it reaches the other extreme end of its trajectory. Now the bowling ball comes swinging back toward my chin. If I have any sense of what conservation means, I have taken great pains not to move *at all* during this demonstration. The goal is to get a least three gasps out of the audience as the ball comes back up, rushing toward my chin, sure to knock me into the next county. The gasps often occur, but the ball never hits me. At best, the bowling ball comes within a whisker of my chin, stops for an instant, and then proceeds to its next cycle. I try my best not to flinch, because though I have complete confidence in the conservation of energy, my mind races a little as the wrecking ball approaches. Did I accidentally give the ball a little push when I let it go? Have I leaned forward a little without noticing it? To this day, neither one has happened, and the bowling ball has never smacked me in the chin. That is a pattern in nature which has immense implications, and not canceling physics classes is one of them!

When we approached linear momentum conservation, the starting point was to multiply both sides of Newton's Second Law equation by the time "t", and then proceed to the definition of momentum as mass multiplied by velocity. The main reason for doing so was that the end product, linear momentum, is conserved in all processes. Rather than multiply both sides by time, we are going to take an alternate approach and multiply both sides of $\mathbf{F_{net}} = \mathbf{ma}$ by a displacement "Δx" and see what unfolds from there. We start with: $\mathbf{F_{net}}(\Delta x) = \mathbf{ma}(\Delta x)$. A couple of details are apparent at the outset. First, we are multiplying vectors again, but this time we are going to use what is termed the "dot product" or "scalar product" to evaluate the result. Second, we are going to assume that all the vectors are in the same direction. The scalar product of two vectors, call them $\mathbf{A}$ and $\mathbf{B}$, is equal to $A \cdot B \cdot \cos ø$, where the angle $ø$ is the angle formed between the two vectors. Therefore, if the vectors are in the same direction, the angle between them is zero, and the cosine of zero degrees is one. This will reduce our result to a scalar product that looks like this: $F_{net}(\Delta x) = ma(\Delta x)$.

Now we combine the only two kinematics formulas we need for motion with constant acceleration, and that were listed pages ago, and in doing so eliminate the variable "t". The simplest way to do this is to use the definition of acceleration, and get the time variable "t" on one side of that equation: $t = (v_f - v_0)/a$. Now since $(\Delta x) = x - x_o$, the other kinematics equation becomes $(\Delta x) = v_o t + \frac{1}{2}(a)t^2$. The next step is to substitute the expression we have for "t" into this equation above and simplify. Give it a whirl – it's a good exercise in basic algebra! If done correctly, you should end up with this result: $a(\Delta x) = (v_f^2 - v_o^2)/2$. The final step is to substitute the expression we obtained for $a(\Delta x)$ back into the $F_{net}(\Delta x) = ma(\Delta x)$ equation we developed by multiplying Newton's Second Law by Δx on both side. The result we end up with becomes vitally important in physics: $F_{net}(\Delta x) = \frac{1}{2} mv_f^2 - \frac{1}{2} mv_o^2$. The term on the left is called the *net work done* on an object, and the $\frac{1}{2} mv^2$ represents the energy of motion of an object, called *kinetic energy*. In words, what we have shown here is that if there is a net force acting on an object that is in the same direction (or opposite direction, since cosine 180 degrees is -1) as the displacement of the object, it results in a change of the object's kinetic energy.

In general then, the work done on an object is defined as the scalar or dot product of the force acting on the object times the displacement of the object times the cosine of the angle between this force and displacement. In equation form: $W = F(\Delta x)(\cos ø)$, where W is the work done and the other symbols are already defined. As defined in *physics*, there are two ways for zero work to be done, even when there is a force acting on an object. One way is for the displacement (Δx) to be zero, such as pushing on a wall. For the sake of clarity, you are doing biological work within your body, but if the wall doesn't move, then no work is being done on *it*. The second way is far more interesting, and that is the case where the force acting on the object is always at right angles to the displacement of the object. Since the cosine of 90 degrees = 0, no work is done in such a situation. We know of such a circumstance, as we have studied it in detail already in the book: uniform circular motion! If an object is moving at constant speed in a circle, then the centripetal force causing that motion, whether it be gravitational, magnetic, or the tension (EM) force in a string, is doing zero work on the object. This makes sense from an internal logic point of view: if zero net work is done, then the kinetic energy of the object cannot change!

Now we take a closer look at the kinetic energy (KE) relation: $\frac{1}{2} mv^2$. First, we should notice that when we double the speed of an object, we get *four* times the KE. Remember that fact the next time you drive a car. Three times the speed implies *nine* times the KE, which in turn implies *nine* times the required stopping distance, assuming we hit the brakes the same way every time, and ignoring the distance traveled during reaction time! It is instructive to contrast KE with momentum, because the two terms are often conflated and confusion ensues. The first difference that emerges is a big one, and that is the fact that KE is a scalar quantity while momentum is a vector quantity. Therefore, when we discuss energy, direction does not enter into the discussion or the resulting equations. Second, it should be clear that if an object has a non-zero momentum, then it must also have a non-zero KE. In brief, if it's moving, it has both. It does *not* follow, however, that if an object has zero momentum, then it must have zero *energy*, because there are other forms of energy besides KE. The only conclusion we can make in such cases is that the object has zero *KE*. This is a common fallacy in logic: if A implies B, then B does not necessarily imply A. One example serves to prove the point. If it rains, I will bring an umbrella. If I bring an umbrella, it may or may not rain! In fact, usually when I bring an umbrella out on the golf course, the Sun comes out and nature gets a good laugh at me.

It also is helpful to distinguish between KE and momentum by looking at things from a quantitative perspective. Suppose two objects have the same momentum, but significantly different masses. Let's make this more concrete: a ball of mass 1 kilogram moving at 100 meters per second, and a human being of mass 100 kilograms, chugging along at 1 meter/second. Both objects have the same momentum, namely 100 metric units (kg•m/s), since **p** = m**v**. How does the KE of the ball compare to the person? Using the formula, KE = ½ mv² derived from Newton's Second Law, we have for the ball: KE = ½ (1)(100)² = 5000 kg•m²/s². You probably have guessed by now, and may have (hopefully) come across the name we use to shorten the energy unit: Joules. This is in honor of James Joule, who we will meet shortly because he plays a critical role in developing energy conservation. The ball has 5000 Joules of KE. Let's look at the person: KE = ½(100)(1)² = 50 Joules! The ball has 100 times more energy than the person, even though they have the same momentum. No wonder the ball hurts a lot more, if it hits us, than if the person does. If the person bumps into us, it's annoying. If the ball hits us, it may kill us.

There are a couple more items to consider before moving on to other forms of energy besides KE. What about the energy required to spin an object? This is called rotational KE, and predictably its formula follows directly from the linear to rotational analog model we have developed for kinematics and dynamics formulas. Instead of the mass "m", we use the distribution of mass "I" (moment of inertia), and instead of the linear speed "v", we use the angular speed ω. Thus, rotational KE = ½ $I\omega^2$. Remember that as long as the angular speed is measured in radians per second, we can connect the linear speed to the angular speed by using the equation of constraint: v = ωR, where R is the radius of the circle being described. The second item requires a good deal of thought and mathematics; I wonder if you have caught the problem after thinking about the basic concepts of space, time, mass, and charge.

The formula for KE is ½ mv^2. The problem with this "classical", or Newtonian description, is that it assumes mass is absolute and constant in all reference frames. We have already seen from STR physics that as the speed of an object increases, its mass, as measured by a stationary observer, increases also. One way to see this stems from the fact that the speed of light is the absolute speed limit of the universe, so if we keep pouring energy into an object, the energy has to manifest in its mass since its speed has an upper limit. The energy has nowhere else to go! We discussed this in the section on "mass", but now we will bring in the equation that predicts what the mass of an object is at any speed "v": $m = m_0/(1 - v^2/c^2)^{1/2}$.

Here, m is the relativistic mass, m_o is the rest mass, and "c" is the speed of light of course. Notice that at low speeds where "v" is much less than "c", the denominator reduces to one, and we have the classical, Newtonian idea that mass is absolute: $m = m_o$. Once again, we should not say that Newtonian Mechanics is wrong, since it works extremely well in literally billions of practical applications. Yet there are those three limitations we have discussed where Newton's Second Law does not work, and speeds approaching the speed of light is one of them. (Recall the other two are inside the atom and at temperatures approaching absolute zero). We invoke the "correspondence principle" for all new theories, forcing them to encompass the correct results predicted by older theories by reducing to those results under specified circumstances, while also presenting new, experimentally proven knowledge from predictions that may not have been envisioned by the older theories. And so, at places like CERN, we accelerate protons and anti-protons in opposite directions around miles of tunnel underground, and eventually they gain speeds that are very near the speed of light. *Every time we do this, the relativistic expression for mass is verified experimentally*. We then smash these proton beams into each other at 99.99% the speed of light, and a vast array of different particles are detected in the shower of fragments resulting from the collisions.

What is the correct formula for KE then, if the ½ mv^2 does not work at high speeds? Though the derivation is not terribly complex, and for you calculus folks not a large stretch at all, it is beyond what the goals of this book aim for, so I will start by asking the reader to picture a graph where the KE of an object is on the Y-axis, plotted against the ratio v/c on the X-axis. Experimental results indicate that the amount of KE increases far more rapidly than the ½ mv^2 prediction from Newton when the speed of the object exceeds even 10% the speed of light. We have always said experiment provides the answer, so whether we like it or not, the Newtonian expression does not work at all speeds. The correct formula for KE, the one that works at *all speeds,* looks like this: $KE = [(1/(1 - v^2/c^2)^{1/2}) - 1]m_o c^2$. This formula exactly matches experimental results. When we use a bit of mathematics called the binomial expansion theorem, we can show that at speeds less than 10% the speed of light, $1/(1 - v^2/c^2)^{1/2} \approx 1 + \frac{1}{2} v^2/c^2$. Plug this back into the relativistic expression and we have the "old" $\frac{1}{2} m_o v^2$ expression: this is the correspondence principle" in action! The upshot of all of this in terms of problem solving and application is as follows: for speeds less than 10% the speed of light, using $KE = \frac{1}{2} mv^2$ will produce extremely tiny, in most cases unmeasurable, discrepancies. Another, equivalent way in which to write the KE expression is a little easier on the eyes, but hides some of the subtleties inherent in the previous formula. The alternate version looks like this: $KE = mc^2 - m_o c^2$, where "m" is the relativistic mass and m_o is called the "rest mass". With this version, we can see that the total energy of a moving object is mc^2, and it is the sum of the rest energy $m_o c^2$ plus the KE of the object. This gives us a far more rigorous treatment as to the origins of the famous $E = mc^2$ equation than the one we heuristically argued our way to through using de Broglie and Planck.

We have developed two fundamental ways to define the concept of energy to this point, calling them "work" and KE. The other fundamental type of energy is called potential energy, or the energy stored in a system. This is how an object can have energy but zero momentum – even though it is not moving, it has potential energy (PE) because it has mass, and also because there may be other stored energy in the system. Each of the four fundamental forces we have encountered, and that exist in our universe as drivers of all the processes we observe, has an associated potential energy linked to it. There is gravitational potential energy for an object existing in a gravitational field produced by mass, there is EM potential energy in the presence of electric charge, and there is nuclear potential energy where there are nucleons present. How do we arrive at formulas for these types of PE?

We have discovered that force can be defined as the rate of change of momentum. This implies that if we were to plot force on the Y-axis, and time on the X-axis, the *area* underneath that graph would be the impulse delivered, or equivalently the change in momentum. We take the same approach when considering PE, except this time we will plot force on the Y-axis versus *distance* on the X-axis. Now the area underneath this graph will be the work done (force x distance) which can represent the amount of energy stored in the system. One example will illustrate this idea in fine fashion: a spring that obeys Hooke's Law. This is the same Hooke who famously had many run-ins with Newton. He showed that many springs are linear, in the sense that if we double the force exerted on the end of the spring, then it stretches double the amount, and this linear relationship holds for a range of possible forces. Not all springs act this way (the slinky would be one notable exception), but many do. In the form of an equation, Hooke's Law looks like this: $F = -kx$. The negative sign is a function of the fact that force is a vector, and it is inserted to show that the force on an object vibrating on a spring always acts toward the equilibrium position where the mass was at rest initially. For our purposes, we can ignore the negative sign and see what a plot of F versus x will bring us. Clearly, the plot is that of a line of slope "k". If we take any point on that line and drop a perpendicular down to the X-axis from that point, we see that there is a triangle "underneath" the diagonal line. The area of that triangle is the work done, which because it is a triangle will be half the base times the height. Hence, the area = $(½ x) \cdot F = ½ x \cdot kx = ½ kx^2$. We now have an expression for the PE stored in a spring! For those schooled in some basic calculus, note that if we take the first derivative of the PE with respect to the changing distance x, we get the force equation. This holds for all types of PE, and should come as no surprise, since the area under a graph of a function is called the integral, and the integral is the anti-derivative of the function! Incidentally, it can be shown that the force within a spring is connected to the EM force at the fundamental, atomic level, and that this potential can work quite well when analyzing the intermolecular forces in the spring.

How does this extend to the gravitational force? We will use the Newtonian conception of gravity here: $F_{gravity} = GMm/R^2$. Plotting the force F versus the distance R produces a function that is clearly not linear, but varies as an inverse square of the distance separating the two objects. Finding the area underneath a curve is not as simple as that of a triangle, so calculus is required here. Ask yourself what function, when differentiated, will produce the expression we have for the force. This is the same as working backward and taking the area or finding the integral of the function. The rule used for differentiating a polynomial function shows that the function we seek must be GMm/R, and that is therefore the PE of a gravitational system. As a detail which in the end helps us conceptually, we know an *attractive* force such as gravity will, when left alone, cause the two masses to come together naturally without any help from an outside agency. This means the PE in a gravitational interaction is given by:
PE = -GMm/R. Another way to "see" this is to think of a ball trapped at the bottom of a potential "hill": in order for it to be liberated from the bottom of this hill, we will need to supply enough energy for that to happen. Hence, it is "bound" in a manner analogous to electrons being bound to a nucleus, which is why all binding energies are listed as negative values.

The electric potential energy is done exactly the same way, since the force between two opposite charges (hence attractive force) is $F_{electric} = kQq/R^2$. Recall that k = Coulomb's constant = $1/4\pi\varepsilon_o$ = 9 x 10^9 Newtons x meters^2/Coulombs^2. Thus, the electric potential energy is given by: PE = -kQq/R. From there it is a short leap to show that the electrical work done in moving a known charge through a voltage difference, since V = voltage = kQ/R, is just the known charge times the voltage difference between the two points in space that locate the start and end points for the charge that is being moved. This explains why the electron-volt is a unit of energy, because 1 eV literally means the amount of energy required to move one electron through a voltage difference of one volt. (That is why 1 eV is not much energy at the macroscopic level, 1.602×10^{-19} Joules, but since electrons and atoms tend to occur in prodigious numbers, a sample of 1 kilogram of anything has a lot of potential energy in its rest mass alone).

Nuclear potential energies tend to be much harder to crack open, mainly because the nuclear force is much harder to describe mathematically. I will leave the formulas out of the mix in this case, and instead emphasize the fact that if we know the force of interaction, we can then find the PE associated with it using the methods already described. Before moving on to the actual conservation law of energy, we will make a quick return to the gravitational PE expression and apply it to objects at or near (a small height above) the Earth's surface. To establish the gain in gravitational PE we get from something by lifting it vertically through the field, we could use the definition of work and reason as follows. The force required to lift an object is at least its weight, mg, and the distance we move it through vertically is the height "h". Since the work done on an object is force multiplied by distance, this seems to indicate that the gravitational PE of an object at or above the surface of the Earth is simply given by PE = mgh. The "zero point" doesn't matter, as it can be chosen anywhere; what *does* matter is the height "h" represents the *change* in the vertical coordinate of the object in the process. Lifting a heavy crate vertically by one meter takes the same amount of energy whether we do it on the first floor of our house or in the attic. So, why do we have to keep saying "at or near the surface" when we use PE = mgh. Look at the formula and I think the reason for this qualifier might emerge.

The problem with this equation is the fact that the value for "g", the acceleration of gravity, is changing while we are lifting the object! The farther away we go from the center of the Earth, the weaker gravity gets. The best way to get at the issue is to return to the graph of force versus distance. In this case, the force of gravity is that inverse square law function created by Newton, and the distance we are looking at starts from the surface of the Earth, a distance R equal to the radius of the Earth since we consider all of the mass of the Earth to be at its center, and ends at R + h. When we go to find the area underneath that small slice of the function, and we stipulate that "h" must be much less than "R" the radius of the Earth, we end up with the PE being very close to the value mgh. We can do this by approximating the area as a rectangle and calculating the PE from there, or by the formal use of integration and calculus. I urge the reader to make a graph and find the area of the rectangle, and remember that the weight of an object is mg, which must equal the pull of the Earth on it: GMm/R^2. For most applications near the surface of the Earth, we can use this simpler version of gravitational PE with extraordinary accuracy. Now that we have the types of energy available, it's time to move on to the actual conservation of *total* energy in a closed system.

It is important to recognize from the outset that it is the *total* energy of an isolated system that is conserved. *The values of KE, PE, and work vary throughout the problem and the process, but the total of all of them must always remain the same, everywhere in the process.* There are so many examples of the power of energy conservation, but I am going to focus on just one, and it is a mechanical system that is a lot of fun: the roller coaster. With all the curved tracks and complicated twists and turns, using dynamics and force diagrams to analyze every point on the track would be next to impossible. Using total energy conservation, finding the speed of the roller coaster at any point is a piece of cake, and knowing the speed at any point informs the design features (such as loops and turns) that need to be incorporated for the ride to be fun and also safe. Let's go for the ride!

The total energy (TE) of a mechanical system can be written as TE = KE + PE + Work. For the coaster car, let's assume it has a little KE at the top of a hill and is about to descend the hill. We want to know the speed of the car at the bottom of the hill. The equation of total energy conservation then looks like this: $\frac{1}{2} mv^2 + mgh + $ Work = constant throughout the process. The "work" factor might be the heat generated by friction in the ride, or the energy used to combat air resistance, or anything external to the system that needs to be accounted for. In most cases, the amount of external work will be small compared to the KE or PE of the system, so we will take it to be roughly zero for this example. Suppose the speed at the top of the ride is 2 m/s, and the car moves down the hill through a height of 40 meters, which is quite a hill! To find the speed at the bottom of the hill, the TE equation then looks like this: $\frac{1}{2} m(2)^2 + mg(40) = \frac{1}{2} mv^2 + mg(0)$. Notice the mass "m" cancels out here, which is most helpful in the design phase for the coaster. We will take g = 10 m/s/s here to make the math simpler to look at. The equation then reduces to: $2 + 400 = \frac{1}{2} v^2$. Solving, the speed at the bottom of the hill, where we take the height to be zero, is the square root of 804, or about 28 m/s, which works out to just over 60 mph!

Not bad, but there are coasters in the world that achieve speeds over 100 mph. If rotational effects need to be considered, then we must add in, as part of the KE portion of the TE, the fact that rotational KE = $\frac{1}{2} I\omega^2$. One can see that, absent frictional effects and other "external" influences, there is a continuous exchange of KE for PE and vice versa during the ride, but the TE must remain the same. The beauty of this approach is that the path taken, unless there are frictional effects, does not matter! All we need to know are the end points, start and finish, and then we plug in the parameters and solve. What occurs between the end points is completely irrelevant in terms of finding the speed at the finishing point chosen. You might have also noticed that nowhere in the TE equations does *time* enter into the picture. Momentum conservation rests squarely on the fact that its validity does not depend on where the experiment is done or how the apparatus is oriented – nature has spatial symmetry. Total energy conservation rests on symmetry in *time*, in that *when* we do the experiment does not matter: the results are the same regardless of our chose time coordinates. Advanced mathematical methods, such as Hamiltonian and Lagrangian dynamics, make use of this symmetry in remarkably effective ways, and if applied judiciously, will show us why that baseball follows the least time path in a gravitational field, known as the familiar parabola.

There is another application of TE conservation that is of terrific use when we study the properties of fluids, which collectively captures both liquids and gasses. It is called Bernoulli's Principle, and his equation looks like this: $P_1 + \frac{1}{2}(\rho)v_1^2 + \rho g h_1 = P_2 + \frac{1}{2}(\rho)v_2^2 + \rho g h_2$. In this equation, P stands for pressure, and the "rho" symbol is the density of the fluid. The subscripts 1 and 2 indicate different locations in the system. I am certain the form of this equation struck you immediately: it looks like the conservation of TE! And in fact, that is exactly what it is based on, just from the perspective of pressure and density. We know by now that every equation tells a story, and Bernoulli's equation, essentially TE conservation, tells a mighty powerful and widely applicable story. If we distill the equation out into words, nature is telling us this: high fluid velocities produce low fluid pressures, and vice versa. Now we proceed to the applications of this principle, derived from TE conservation.

When an artery narrows because of plaque, the speed of the blood flow must increase in that region, just as the water coming out of a hose increases its speed when we taper down the opening. This increased speed means a reduced pressure in that region, giving rise to a pressure difference in the artery that is dangerous and sometimes lethal. When high winds, such as in a tornado, blow nearby a building, there is a low pressure zone created there that often is reported by people saying they felt their eardrums "pop". The normal, higher pressure within the house then exerts a net force outward, causing the roof to explode off the house or typically some other structural failure. As a word of caution, studies have shown that opening windows in an attempt to balance the pressure during a tornado does not work, and actually increases the danger of injury due to flying debris.

The examples of Bernoulli's Principle in action are seen everywhere. The shower curtain comes in toward you when you blast the water: high speed fluid motion in the shower creates a low pressure zone, so the curtain moves to the low pressure area. Airplane flight and lift are deeply reliant on the fact that wings are designed (along with flaps down) to increase the air flow over the top of the wing, producing a lower pressure above the wing than below the wing, which in turn provides lift for the plane to fly. Race car drivers know very well that "drafting" behind a fast moving vehicle can save fuel, because the low pressure zone directly behind the racing car is caused by the fast moving air. Carburetors work off Bernoulli's Principle, hence energy conservation. Since liquid gasoline, like all liquids, does not burn, we need to turn the liquid gasoline into a vapor before injecting it into the cylinder to be ignited by the spark plug or, in a diesel engine, by the high compression caused within the chamber. How do we change a liquid into a gas? We do this either by raising its temperature, or by reducing the pressure around it, or both. In the case of the carburetor, we use tapered nozzles called "Venturi jets" to increase the speed of the gasoline, reducing the pressure, causing the gas to vaporize as it mixes with air before being injected into the combustion chamber. Yet another example is the fact that small children are cautioned not to stand too close to the edge of a train platform for many reasons, but one of them is the fact that high speed trains passing through will produce low pressure zones under the train, causing the risk of people being sucked into the low pressure zone. The applications are virtually endless.

We now return to electrical systems and Kirchhoff's Rules. Recall that there were two rules, and one of them is the "point rule" which is used for currents in a circuit and is based on the conservation of charge. The second rule is called the "loop rule", and it states that around any closed loop in a circuit, the voltage drops and rises must add up to zero. This is a statement of energy conservation in an electrical system, since voltage is energy divided by charge. These two rules work in tandem to enable physicists and engineers to solve any circuit, no matter how complex, because the rules are based on conservation laws.

The last item we will look at in this section, and the one to close this journey and the book, is the field of thermodynamics. Thermodynamics literally means "heat in motion". This is the field where the conservation of total energy had its birth as a concept, and the ensuing decades only served to greatly amplify the importance and usefulness of all the conservation laws. We arrive at this point with the knowledge of Newton's Second Law, the conservation of linear momentum, and the definition of pressure as force divided by area. Boltzmann showed with statistical mechanics that the average KE of molecules in a gas depends only on the temperature of the gas; in equation form, $KE_{average} = (3/2)kT$, where k = Boltzmann's constant and T is the temperature of the gas in Kelvin. From these ideas and some assumptions, the ideal gas law can be derived: $PV = nRT = NkT$. Historically, a puzzle emerged in the study of gasses that James Joule and many others worked on for years. We will turn to that puzzle now, and see how it was resolved.

Picture a bicycle pump, which is basically a cylinder, filled with a gas (air), and equipped with a movable piston. This is the system we will use throughout the discussion which follows. Experimental results consistently show that it takes far more energy to raise the temperature of a gas at constant pressure versus at constant volume. Using the bicycle pump as the model, if we want to triple the temperature of the air inside the cylinder, we start by supplying heat. If we hold the handle of the pump so that the volume of the air inside does not change, a process called isochoric, then the amount of heat required to raise the temperature of the air is far less than if we repeat the process but let go of the handle so that the *pressure* remains constant throughout the (isobaric) process. It took James Joule to see that the reason for this is directly related to energy: in the isobaric process, work is being done to move the piston up the cylinder wall – the gas is expanding. This takes energy to do, so work must be a form of energy! We are tacitly assuming here that one cannot "get something for nothing". And that is the birth of total energy conservation, which formally states that the total amount of energy in a closed system remains constant.

We will dig deeper now, and uncover the three laws of thermodynamics, and in the process develop an understanding of two and four-stroke heat engines, efficiency, and much more. The place to start is to define heat energy, and this is done by first recognizing that heat *does not* equate to temperature, at least not in the field of physics. Heat is represented by the symbol Q, and it involves a transfer of energy by either conduction, convection, or radiation. Conduction transfer is principally for solids, and involves moving heat down a chain of fixed molecules or atoms. One example would be holding a nail in a candle flame. Eventually, the heat from the candle will reach your hand by conduction, even though the molecules and atoms in the solid are not physically translating through space. Convection is heat transfer in fluids, and occurs by the mass movement of molecules. One example is hot air rising. Radiation heat transfer is the transfer of heat energy by EM waves, mainly in the infrared region of the spectrum. Microwave ovens are an example of this type of heat transfer.

When we delve into heat energy, we invariably come across an idea called "specific heat capacity". This refers to a specific substance, and is defined as the amount of heat required to raise a given mass of the substance by one degree Celsius or Kelvin. Because of the molecular structure of sand, it has a very low specific heat capacity. This means sand gets hot very fast on a summer's day, and cools down rapidly as well. Water, on the other hand, has a very high specific heat capacity, which is what makes it so useful to store energy for long periods of time. When you go to the beach in the summer, there is nearly always a breeze blowing toward the shore during the daytime, because the low specific heat of the sand causes it to get hot, and hot air rises. Once the hot air starts to rise, the cooler air over the ocean rushes in to take its place! At night, the process is reversed, and ocean breezes tend to blow out to sea during those moonlit walks. There are exceptions to this, but in general the observation holds true.

The formula for heat transfer that we use in thermodynamics is $Q = mc\Delta T$, and in this case "c" is *not* that "other thing" – here it stands for specific heat capacity! The other symbols are already defined. However, when it comes to a gas inside a bicycle pump or any heat engine, we employ the use of moles instead of mass, and so the formula becomes $Q = nc\Delta T$. This is where we discovered that the specific heat capacity at constant pressure, with symbol c_p, is *always* greater than the specific heat capacity of the same gas at constant volume, c_v. Symbolically, $c_p > c_v$. It turns out that for a monatomic gas, $c_p = 5/2(R)$ and $c_v = 3/2(R)$, where R is the gas constant from the ideal gas law. For a diatomic gas, it is shown experimentally that

$c_p = 7/2(R)$ and $c_v = 5/2(R)$. The story being told there is very logical: it takes more heat to raise the temperature of diatomic gasses because those molecules have more degrees of freedom to absorb the heat without moving faster. Monatomic gas molecules have no choice but to move faster when heat is supplied, whereas diatomic molecules can use the heat to rotate or vibrate as well as translate through space. As we have seen, temperature measures the average translational KE of gas molecules, so diatomic gasses take more heat energy than monatomic gasses to increase their temperature and translate through space faster.

The internal energy of a gas, such as the air trapped inside the bicycle pump cylinder, depends only on temperature, and hence only on the average KE of the gas molecules. The symbol we use for internal energy is typically U, and so the formula for its energy content is given by $\Delta U = nc_v\Delta T$. Curiously enough, this equation holds for all processes in thermodynamics, whether it is a constant volume (isochoric) process or not. It appears internal energy change relies solely on temperature changes in the gas, and not changes in pressure or volume. We are now ready to put this all together and explore the three laws of thermodynamics.

We need to develop sign conventions for the change in internal energy of the gas, the heat transfer, and the work being done in the process. Once the conventions are established, we can express the conservation of total energy in equation form as applied to thermodynamic systems. The conventions are logical: if the temperature of the gas increases, ΔU is positive, if heat energy goes *into* the gas, Q is positive, and if work is done *by* the gas so the volume expands, then W is positive. Now the first law of thermodynamics, which holds that the total energy of a closed system remains constant throughout the process, can be written in the following way: $Q = \Delta U + W$. What could be more simple, and yet more profound? If we supply heat to the air inside that bicycle pump, then either the internal energy of the gas will increase, or work will be done by the gas and the piston will move up the cylinder increasing the volume the gas occupies, or both. We are accounting for all the energy transfer, and we are essentially saying "you cannot get something for nothing". There is no way a closed system can produce more energy than it takes in.

There are four basic processes that we look at in thermodynamics, and we have discussed two of them already: isobaric (constant pressure) and isochoric (constant volume). The other two are called isothermal (constant temperature) and adiabatic (zero heat transfer). Let's begin with the adiabatic process, a word that originates from Greece and roughly translates to "no passage through". Suppose we compress the air in the bicycle pump by pushing down on the handle, and we do so very swiftly so no heat is allowed in or out of the cylinder in the process. The other way this can be accomplished is by insulating the cylinder, much like what is done for a thermos bottle. Since the process is adiabatic, $Q = 0$ by definition. The work done here must be negative, since we are compressing the gas, not expanding it. This means, using the first law of thermodynamics, that the internal energy change ΔU must be positive, and thus the temperature of the gas must increase! Of course that is what every child knows: when you pump air into a tire, the pump gets hot!

Now let's try to expand the gas, the air in the cylinder, using an isothermal process. Since this means, by definition, that we are not allowing the temperature of the gas to change, we have $\Delta U = 0$. The average kinetic energy of the air molecules cannot change. How can we make the gas expand, increasing its volume, and thus do work without changing its temperature? The first law of thermodynamics provides the solution: use *all* the heat supplied into the gas to do work by moving the piston up the cylinder and increase its volume without changing its temperature. The first law then looks like: $Q = 0 + W = W$. The heat Q is positive because it is into the gas, and the work is positive because the volume is increasing and work is being done by the gas. The first law is as logical as it is predictive, and it all goes back to total energy conservation. All four of these processes can be graphed by putting pressure on the Y-axis and volume on the X-axis, but for now we will move on to a more pressing matter. The first law does not prohibit a 100% efficient process, where the work we get out of a heat engine is equal to the heat we put into the gas during the cycle. Yet we have never engineered or observed a 100% efficient system of any kind, whether it be mechanical, electrical, or thermodynamic. Why?

The problem, of course, is that the universe appears to dramatically favor processes that are not reversible. When we drop an egg onto the floor, the law of total energy conservation does not prevent the egg from reassembling and popping back up into our hand. But that never happens, at least it hasn't for me! We have encountered this issue before, along with our friend Ludwig Boltzmann, and it is this idea of entropy that has gummed up the works. Recall that entropy is a measure of the randomness in a system, and that maximum entropy occurs when the system reaches equilibrium. This is why the sugar cube dissolves in your coffee and spreads out randomly, yet never do we see the granules come back together into a perfect cube and then pop out of the coffee into your hand. If you observe that happening, the vast probability is that you are not drinking coffee! The second law of thermodynamics has several alternate expressions, but one of them simply states that natural processes proceed along paths that have more entropy, because those paths are more probable. When it comes to heat engines, this means that no matter what we do, some of the heat we put into the engine will be "wasted" and therefore not used to do work. While the total energy of the system is constant, entropy content is not constant: it increases because that is the probability of what will happen. This is why we should never call it an "energy crisis", since the total amount of energy in the universe is the same as it always has been and always will be (assuming our universe is a closed system). Our energy policies should focus more on the "entropy crisis". It isn't the *amount* of energy that is at stake here, it is the *type* of energy available for us to do the work that is vitally important. Since heat energy has the most entropy, we would like to minimize wasted heat in a closed system process.

Another way to state the second law of thermodynamics is to posit that no heat engine, operating in a cycle, can be 100% efficient. There will always be some waste heat involved. In the typical internal combustion engine, there are four strokes: input the fuel-air mixture, compress the mixture in the cylinder with a piston, explode the mixture (either with a spark plug or by the high compression ratio of a diesel engine) in a power stroke so that the piston slams back up the cylinder wall (taking the connecting rod to the crankshaft with it), and finally the exhaust stroke where the residue not used in the cylinder is blasted out the tailpipe. The general formula for the efficiency of a heat engine, such as the one in all internal combustion engines, is the net work done by the gas during the cycle divided by the heat put into the gas during the cycle, or output work divided by input heat. A very clever French engineer, Sadi Carnot, showed that the most efficient four-stroke cycle for a heat engine would be isotherm-adiabatic-isotherm-adiabat. This maximum efficiency is given by the formula $\text{eff}_{max} = 1 - T_{cold}/T_{hot}$, where the temperatures must be in Kelvin, and the temperatures represent the coldest and the hottest extremes reached by the gas during the cycle. To engineer a Carnot engine is exceedingly challenging, and no-one has done it to date that I am aware of, but we certainly could do much better. The typical internal combustion engine in a car today is around 20% efficient, which means for every dollar of gasoline purchased, 80 cents goes out the tailpipe. Obviously higher efficiencies mean greater gas mileage, so you can see that there is a counter-incentive at work here, because better mileage means less profit for the oil industry.

The second law is often paraphrased by saying "you can't break even", because 100% efficiency is effectively impossible. I remember reading somewhere that the probability of that egg re-assembling and jumping back into your hand is about the same as a chimpanzee randomly typing on a computer keyboard and coming up with the complete works of Shakespeare. How on Earth anyone ever calculated that is beyond me, but it does paint a picture that accurately shows the meaning etched on Ludwig's tombstone: in words, entropy is proportional to probability.

Maybe you have already seen the loophole here that must be closed, and that leads to the third law of thermodynamics. If we allow the coldest temperature to be zero degrees Kelvin, then according to Carnot, 100% efficiency is achieved! Check the formula out and verify this for yourself. Well, that means the third law of thermodynamics is going to tell us that we cannot reach zero degrees Kelvin, known as absolute zero. Apparently, we cannot get something for nothing, we can't break even, and now we cannot get out of the game. One can see that there is another problem with absolute zero, and that connects back to the HUP. According to kinetic theory, if the temperature is zero, then the average KE per molecules is zero. But that would mean we know the position and the momentum of the gas molecules simultaneously to an unlimited degree of accuracy, and that violates the Heisenberg Uncertainty Principle, which undergirds all of quantum physics. The best we have done to date is a tiny fraction of a degree about absolute zero, but it's still not zero.

We have come to the end of this journey, over 100,000 words long, and yet dramatically shorter than the giant textbooks people tote around these days. I kept my promise on that front! I hope you will take with you an abiding sense of wonder regarding the things we know and understand about our universe, but perhaps more importantly, you will have a deeper appreciation for all the things we still do *not* know and have yet to discover. Why is the expansion of the universe accelerating? What constitutes 95% of the mass of the universe – what is dark energy and what is dark matter? Why is the universe made in such a way that we move toward more entropy, all through time? What is the ultimate fate of the universe? And the best one of all: how do we explain this fact of our existence itself? Atoms combining in a way so as to form life, a life and a brain made in a way that can contemplate its own walk between the two eternities that serve as endless bookends to all the sorrows and joys we experience as living human beings here on Earth. Life is a miracle and a wonder, and a gift to be treasured.

Acknowledgements

I may be breaking precedent by writing another type of dedication at the *end* of a book; if true, then so be it. This book is a distillation of over forty years of teaching students in many settings, and yet more accurately, it is a summary, far from complete, of what I have *learned* along the way. It springs from memories of things I have gleaned through using dozens of textbooks, reading thousands of articles, and sending out thoughts verbally in the hope that my explanations are bringing deeper understanding. The main source that gave life to these words are the questions my students have raised over the years, questions that have caused me to think more deeply and express myself more clearly. Letters and conversations from so many former (and current) students remain the lasting paychecks in the teaching profession.

It has been my great fortune to have had some master teachers and coaches along the way. I think of Miss Katie Reardon, my high school math teacher, who put me on the path to understanding the beauty and power of mathematics. I think of several great coaches I have worked with in football, basketball, and baseball, all of whom showed me how much teamwork and collaboration can accomplish through hard work, discipline, and expert instruction. Coach Thomas Blackburn comes to mind; if you have seen *Hoosiers*, you have met Tom. My good friend Joel Strogoff showed me, by example, how to lead an institution with fairness, integrity and vision. George Andes, one of the most brilliant individuals I have known in my life, and also the kindest of souls, was a fellow physicist who was a joy to be around. I recall many talks we had about all sorts of ideas in physics, and without fail his enthusiasm and his knowledge were inspirational. All of these people, and many more colleagues and students too numerous to list here, may not be well known to all of humanity, but they changed many lives for the better on a daily basis.

Perhaps the greatest inspiration of them all has been my 104 year-old mother. Her masterpiece is the way she lives her life: always open to ideas different from her own, full of good (mildly sarcastic) humor, and infinitely kind. She is the living example of the golden rule, and she teaches everyone around her, by her actions more than her words, to walk a mile or more in another's moccasins, and even beyond that mile, to not gossip or judge. Finally, my daughter Sarah, the greatest gift in my life, and my entire family, have created that safe harbor from all the inevitable storms in life. On that front, too, I hit the lottery.

George C. Whittemore – July 4, 2020

Princeton, Massachusetts

All Rights Reserved

Copyright 2020

www.ingramcontent.com/pod-product-compliance
Lightning Source LLC
Chambersburg PA
CBHW080451220526
45465CB00006B/2238